THE RIGHT

T0073378

The right to clean water has been adopted by the United Nations as a basic human right. Yet how such universal calls for a right to water are understood, negotiated, experienced and struggled over remain key challenges. This book elucidates how universal calls for rights articulate with local historical geographical contexts, governance, politics and social struggles, thereby highlighting the challenges and the possibilities that exist. Bringing together a unique range of academics, policy-makers and activists, the book analyzes how struggles for the right to water have attempted to translate moral arguments over access to safe water into workable claims. This book is an intervention at a crucial moment into the shape and future direction of struggles for the right to water in a range of political, geographic and socio-economics contexts, seeking to be pro-active in defining what this struggle *could* mean and how it might be taken forward in a far broader transformative politics. The book engages with a range of approaches that focus on philosophical, legal and governance perspectives before seeking to apply these more abstract arguments to an array of concrete struggles and case studies. In so doing, the book builds on empirical examples from Africa, Asia, Oceania, Latin America, the Middle East, North America and the European Union.

Farhana Sultana is Assistant Professor of Geography at the Maxwell School of Syracuse University, USA. Her research interests and publications are in water governance, political ecology, gender and development. Combining insights and experiences in and outside academia, she engages in critical interdisciplinary research on water in the global South.

Alex Loftus is Lecturer in Geography at Royal Holloway, University of London, UK. His research focuses on the political ecology of water and the political possibilities within urban ecologies. He is the author of *Everyday Environmentalism: Creating an Urban Political Ecology* (University of Minnesota Press).

THE RIGHT TO WATER

Politics, governance and social struggles

edited by

*Farhana Sultana and
Alex Loftus*

publishing for a sustainable future

LONDON AND NEW YORK

First edition published 2012
by Earthscan
2 Park Square, Milton Park, Abingdon, Oxon OX14 4RN

Simultaneously published in the USA and Canada
by Earthscan
711 Third Avenue, New York, NY 10017

Earthscan is an imprint of the Taylor & Francis Group, an informa business

British Library Cataloguing in Publication Data
A catalogue record for this book is available from the British Library

Library of Congress Cataloging in Publication Data
The right to water : politics, governance and social struggles /
[edited by] Farhana Sultana and Alex Loftus. – 1st ed.
p. cm.
Includes bibliographical references and index.
1. Right to water. I. Sultana, Farhana. II. Loftus, Alex.
K3260.R54 2010
346.04′32–dc23
2011024261

ISBN: 978-1-84971-360-3 (hbk)
ISBN: 978-1-84971-359-7 (pbk)
ISBN: 978-0-203-15210-2 (ebk)

Typeset in Times NR MT
by Graphicraft Limited, Hong Kong

CONTENTS

CONTENTS

FIGURES

TABLES

CONTRIBUTORS

Editors

Farhana Sultana is Assistant Professor of Geography at the Maxwell School of Syracuse University, USA. Her research interests and publications are in water governance, political ecology, gender and development. Combining insights and experiences in and outside academia, she engages in critical interdisciplinary research on water in the global South.

Alex Loftus is Lecturer in Geography at Royal Holloway, University of London, UK. His research focuses on the political ecology of water and the political possibilities within urban ecologies. He is the author of *Everyday Environmentalism: Creating an Urban Political Ecology* (University of Minnesota Press).

Contributors

Thomas Appleby is Senior Lecturer in Law at the University of the West of England, Bristol, UK. His current research focuses on marine environmental regulation and has had significant practical application in the UK, the EU and around the world.

Karen Bakker is Associate Professor (Geography) and Director of the Program on Water Governance, at University of British Columbia, Vancouver, Canada. Her research interests span political ecology, political economy and development studies. Her latest book is *Privatizing Water: Governance Failure and the World's Urban Water Crisis* (Cornell UP, 2010).

Maude Barlow is Chairperson of the Council of Canadians, former Senior Advisor on Water to the 63rd President of the UN General Assembly, and recipient of the 2005 Right Livelihood Award. Her latest book is *Blue Covenant: The Global Water Crisis and the Coming Battle for the Right to Water* (New Press, 2008).

Patrick Bond is a political economist at the University of KwaZulu-Natal in Durban, South Africa, where he directs the Centre for Civil Society (http://ccs.ukzn.ac.za). Amongst his books are *Politics of Climate Change* (2011); *Looting Africa* (2006); *Talk Left, Walk Right* (2006); *Elite Transition* (2005); and, *Unsustainable South Africa* (2002).

Rocio Bustamante is Assistant Professor and Senior Researcher at the Andean Centre for Water Use and Water Management (Centro AGUA) at San Simon University (UMSS), Cochabamba, Bolivia. Her background is in Law and Political Sciences, and her research is concerned with water governance and legal pluralism.

Krista Bywater is Assistant Professor of Sociology in the Department of Sociology and Anthropology at Muhlenberg College, USA. She was an Andrew W. Mellon Postdoctoral Fellow at Grinnell College and a Dissertation Fellow at Skidmore College. Her areas of specialization are development studies, environmental sociology and globalization studies.

Cristy Clark is a PhD candidate with the Australian Human Rights Centre in the Faculty of Law at the University of New South Wales, Sydney, Australia. Her research focuses on the prerequisites for realizing the right to water for the urban poor and includes case studies in Manila and Johannesburg.

Carlos Crespo is a Sociology Professor and Researcher at San Simon University (UMSS), Cochabamba, Bolivia, and the head of the Environmental Studies Section at the Centre for Advanced Studies (CESU). His research focuses on the political ecology of water and other natural resources, spatial segregation, racism, anarchy and autonomy.

Ilaria Giglioli is a PhD student in the Geography Department, University of California Berkeley, USA. Her research and professional work has focused on water politics in Sicily and Palestine. She has published in the *International Journal of Urban and Regional Research* and in *Conflitti Globali*.

Evadne Grant is Principal Lecturer in Law and Director of Law Postgraduate Programmes in the Law Department, University of the West of England, UK. Her research focuses on ways of balancing the sustainable implementation of the right to food, water and housing with respect for the environment and human dignity.

Andrea Keessen is Assistant Professor at the Institute for Constitutional and Administrative Law, Centre for Environmental Law and Policy, and Faculty of Law, Economics and Governance at Utrecht University, Utrecht, Netherlands.

Jamie Linton teaches in the Department of Geography at Queen's University, Canada. His research focuses on the history and culture of water and its relevance to water governance. His latest book is *What is Water? The History of a Modern Abstraction* (UBC Press, 2010).

Katharine Meehan is Assistant Professor of Geography at the University of Oregon, USA. Her research focuses on water informality as a mode of governance and development in the urban global South, particularly Mexico.

Kyle R. Mitchell is a PhD candidate in Sociology at the University of Strathclyde in Glasgow, Scotland. His research interests include the political economy of water, specifically, and the broader topics of neoliberal environments, the idea of human rights and property relations.

Verónica Perera is Assistant Professor of Sociology at SUNY Purchase in New York, USA. Her publications and research interests focus on neoliberalism and transnational social movements, Latin American water struggles, and the embodiment of activism.

Jacinta Ruru researches the legal rights of Indigenous peoples to own and govern land and water at the University of Otago, Aotearoa New Zealand. She guest edited a special issue of the *Journal of Water Law* titled 'Contemporary Indigenous Peoples' Legal Rights to Water in the Americas and Australasia' (2010).

Jeremy J. Schmidt is a Trudeau Scholar and PhD Candidate in the Geography Department, University of Western Ontario, Canada. He is co-editor (with Peter G. Brown) of *Water Ethics: Foundational Readings for Students and Professionals* (Island Press, 2010). His interests are in water governance in Nova Scotia (www.waterethic.ca).

Chad Staddon is Senior Lecturer in Geography and Environmental Management and founder of the Bristol Group for Water Research, an interdisciplinary research group, at the University of the West of England, UK. His research focuses on the social and policy dimensions of water supply, stakeholder involvement and water sector regulation.

Marleen van Rijswick is Professor of Water Law at Utrecht University, Utrecht, Netherlands. She has researched the protection of water quality, environmental and water law in general and on adaptation to climate change. Research projects take the protection of citizens and common pool resources as a starting point.

Anna Maria Walnycki is an anthropologist undertaking doctoral research at The STEPS Centre at The Institute of Development Studies, UK, where she is exploring the development and the institutionalization of 'The Right to Water for Life' Bolivia, and its implications for providers and communities in peri-urban areas.

PREFACE

This book first began to germinate in a set of discussions the two of us shared around the water justice movement whilst working in separate Geography departments in the University of London system. Our hopes of developing a forum through which critical dialogue around the right to water could be instigated, worked through and developed, were briefly curtailed when Farhana moved across the pond to Syracuse University. Nevertheless, when an opportunity arose, we were quick to return to the plans and were immensely fortunate in being able to secure funding to bring together a range of scholars and activists in a two-day event at the Maxwell School of Syracuse University. This international conference on 'The Right to Water' took place over March 29 and 30, 2010, beginning with a series of keynote speeches by Patrick Bond, Bill Derman, David Getches, Anil Naidoo, Darcey O'Callaghan and Oren Lyons, and continuing on the following day when two dozen papers were presented and discussed. These papers were structured around thematic clusters: philosophical perspectives, legal perspectives, governance and social struggles. Some, in revised form, are included in this book, in addition to new contributions from scholars who were invited to submit chapters.

Above all, our goal in the conference was to create space for dialogue and debate among scholars, activists and practitioners. This space then became an interdisciplinary and international platform from which strategic possibilities for ensuring equitable access, use and availability of water worldwide began to be worked through. Some of the key questions that were addressed included: How important is the human right to water – and how is it mobilized – in different struggles for equitable access to water? How influential are international discourses on rights in shaping access to water in different contexts? How do broader discourses articulate with local historical geographies of struggles for water and rights discourses? Such questions inform and animate this book, which consists of a selection of contributions across a range of conceptual and practical exemplars. All the chapters resonate with and intervene in debates around the right to water in academic, policy and activist communities. Given the range of backgrounds of the contributors, the political pertinence of the issues being discussed and the effort to define in bold, proactive and provocative ways what struggles for the right to water might become, our hope is that the ideas, insights and problematizations herein will critically advance existing debates, practices and policies in configuring the right to water in more just and equitable ways globally. As we elaborate in our introduction chapter, this is imperative in the contemporary moment and now still timelier given the United Nations' resolutions on the right to water in late 2010. As states, citizens, and groups start to move towards working out the details of national policies and implementation plans, we hope the insights and ideas in this book will be instructive and inspirational.

The book is a product of conversations and collaborations with many scholars, practitioners and activists across space and time, but especially with the twenty-one contributors located across the globe. In undertaking the ambitious goals to have all the chapters written, reviewed, revised and re-submitted within a few months' timeframe, we have been privileged to work with gracious and attentive friends and colleagues. We thank all the chapter contributors for being part of this collective journey, but more importantly, for brilliant expositions and thoughts that make this book a whole. We also thank all the conference presenters and speakers, and the two hundred participants, for excellent debates and thought-provoking discussions, all of which helped inform the book in one way or another.

The conference would not have been possible without the support of sponsors at Syracuse University, whom we would like to acknowledge: Department of Geography, Program for the Advancement of Research on Conflict and Collaboration, Center for Environmental Policy and Administration, Environmental Finance Center, Syracuse Center of Excellence, South Asia Center, Maxwell Dean's Office, College of Arts and Sciences Co-Curricular Grant, Chancellor's Feinstone Grant for Multicultural Initiatives, International Relations Program, Africa Initiative, and Program on Latin America and the Caribbean. We would like to thank the Vice-Chancellor, Dean of the Maxwell School, and Dean of the College of Arts and Sciences at Syracuse University for supporting and participating at the conference. Our gratitude also goes to our graduate students who helped out at the conference or in the preparation of the book, especially Emera Bridger Wilson, Clint Misamore, Sara Bittar and Fiona Nash. Special thanks to Jonathan Chowdhury for support throughout the conference and in the preparation of the book, in particular for his help with the conference poster design and book cover design.

This book was put together during an incredibly busy time for both of us at critical junctures in our academic lives, and we are grateful to all those who supported us along the way, especially our families.

Finally, we would like to thank Tim Hardwick at Earthscan for seeing the book through from the beginning, and to several colleagues at Taylor & Francis for assistance in the production stage.

This book is dedicated to people around the world who continue to struggle for water. May all our futures with water be more equitable and just.

Farhana Sultana and Alex Loftus
Syracuse, NY and London, UK
May 2011

FOREWORD

Maude Barlow

On July 28, 2010, the United Nations General Assembly adopted an historic resolution recognizing the human right to safe and clean drinking water and sanitation as "essential for the full enjoyment of the right to life." For those of us in the balcony of the General Assembly that day, the air was tense with suspense. A number of powerful countries had lined up to oppose it so it had to be put to a vote. Bolivian UN Ambassador Pablo Solon introduced the resolution by reminding the assembly that humans are about two-thirds made of water and our blood flows like a network of rivers to transport nutrients and energy to our bodies. "Water is life," he said.

But then he laid out the tragic and growing numbers of people around the world dying from lack of access to clean water and quoted a new World Health Organization study on diarrhoea showing that every three and a half seconds in the developing world, a child dies of water-borne disease. Ambassador Solon then quietly snapped his fingers three times and held his small finger up for a half second. The General Assembly of the United Nations fell silent. Moments later, it voted overwhelmingly to recognize the human right to water and sanitation. The floor erupted in cheers.

Two months later, the UN Human Rights Council adopted a second resolution affirming that water and sanitation are human rights, adding that the human right to safe drinking water and sanitation is derived from the right to an adequate standard of living and is "inextricably related to the right to the highest attainable standard of physical and mental health as well as the right to life and human dignity." The two resolutions together represent an extraordinary breakthrough in the international struggle for the right to safe clean drinking water and sanitation and a crucial milestone in the fight for water justice. They also complete the promises of the 1992 Rio Earth Summit where water, climate change, biodiversity and desertification were all targeted for action. All but water had been addressed by the United Nations with a convention and a plan; now the circle is closed.

The struggle to achieve this milestone was a long one and blocked for years by some powerful corporations and governments who prefer to view water as a private commodity to be put on the open market for sale. Indeed, forty-one countries, including the UK, Australia, Japan, Canada and the US, abstained in the General Assembly vote (although the US voted in favour of the resolution that came before the Human Rights Council). Some of these governments insist that they are still under no new obligations in this area, as they claim the General Assembly vote was not binding. This is incorrect. Because the Human Rights Council resolution is an interpretation of two existing international treaties, it clarifies that the resolution adopted by the General Assembly is legally binding in international law. Said an official UN press release, "The right to water and sanitation is a human right, equal to all other human rights, which implies that it is justiciable and enforceable."[1]

This means that whether or not they voted for the right to water and sanitation, every member state of the United Nations is now required to prepare a Plan of Action for the Realization of the Right to Water and Sanitation and to report to the UN Committee on Economic, Social and Cultural Rights on its performance in this area. This plan of action must meet three obligations: the *Obligation to Respect*, whereby the state must refrain from any action or policy that interferes with these rights, such as withholding water and wastewater services because of an inability to pay; the *Obligation to Protect*, whereby the state is obliged to prevent third parties from interfering with these rights, such as protecting local communities from pollution and inequitable extraction of water by the private sector; and the *Obligation to Fulfil*, whereby the state is required to adopt any additional measures directed toward the realization of these rights, such as providing water and sanitation services to communities currently without them.

Already, the resolutions have had their first successful test case. The Kalahari Bushmen of Botswana have been fighting for decades to regain access to their ancestral homes in the Kalahari Desert, which they finally won in a Botswana Court in 2006. However, that same court denied them access to their traditional water sources, a borehole the government had smashed several years earlier. The Bushmen appealed that ruling and in a momentous January 2011 decision citing the UN's new recognition of the right to water and sanitation, Botswana's Court of Appeal unanimously quashed the earlier ruling and found that the Bushmen have the right to use their old borehole as well as the right to sink new boreholes and called their treatment by the government "degrading." In its judgment, the Court said it is "entitled to have regard to international consensus on the importance of access to water" and referenced the two UN resolutions.

These historic resolutions present an incredible opportunity for other groups, communities and Indigenous peoples around the world suffering from water shortages, unsafe drinking water and poor or non-existent sanitation services. It is not often that a new right is recognized at the United Nations, especially around an issue as increasingly political and urgent as the global water crisis. The right to water and sanitation are living documents waiting to be used for transformational change around the world. This is why the book you hold in your hands is so important as it explores the issues surrounding the right to water and lays down a challenge to stretch our minds and our policies to set a path toward a water-secure future for all.

Will the right to water and sanitation be defined in the more traditional, "western" notion of rights, what are often referred to as "first generation rights," which exist to protect the individual from excesses of the state, or will it be defined in a more inclusive way, embracing "second" and "third" generation rights more closely related to issues of social and economic equality and even group and collective rights such as those found in the UN Declaration on the Rights of Indigenous Peoples? Will the genuine realization of these new rights require recognizing and honouring that some cultures place responsibility and relationship of community over the more traditional UN definition of individual rights? Will it be possible to protect the human right to water and sanitation without recognizing the inherent rights of nature and other species? Is weaving the rights of nature into the interpretation of the human right to water and sanitation essential for true transformation?

These and other crucial questions lie before us, in the pages of this book and in the work that calls our name. *The Right to Water: Politics, Governance and Social Struggles*, edited by Farhana Sultana and Alex Loftus, is a brilliant collection of essays from the best thinkers, academics and activists in the field, and is required reading for all those wanting this mighty effort to succeed. One thing was clear to me, however, on that warm July day at the UN when the General Assembly voted to recognize the human right to water and sanitation. Every now and then, humanity takes a collective step forward in its evolution as a species. The recognition that no one should have to watch a child die because of an inability to pay for clean water is one such step.

Note

1 October 10, 2010 press release from the Office of the High Commissioner for Human Rights quoting Catarina de Albuquerque, then the Independent Expert on human rights obligations related to access to safe drinking water and sanitation (now the Special Rapporteur), entitled "UN united to make the right to water and sanitation legally binding."

1

THE RIGHT TO WATER

Prospects and possibilities

Farhana Sultana and Alex Loftus

Introduction

Water is life-giving and non-substitutable. Yet safe water remains inaccessible to millions of people around the world. Given this, the fundamental importance of fulfilling people's right to water could not be clearer. Indeed, it is not surprising that calls for the right to clean potable water have galvanized scholars, activists and policy-makers, whilst struggles over this right have emerged as a focal point for political mobilization in a range of locations globally (Gleick, 1999; Petrella, 2001; Barlow and Clarke, 2002; Shiva, 2002; WHO, 2003; UNDP, 2006; Barlow, 2008; Bond, 2008). Global and local movements have highlighted the critical need for water justice, in a world where nearly a billion people still lack safe drinking water and water-related deaths remain the leading cause of infant mortality in the developing world. The relatively modest costs of providing safe potable water and the continuing high rates of illness and death from water-related diseases have resulted in the provision of safe water gaining prominence within the Millennium Development Goals (MDGs). It also formed the crux of a rallying call for water activism for the right to water. Although recognizing the right to water was in part formalized in the UN Committee on Economic, Social and Cultural Rights General Comment No. 15 of 2002, and embodied in the 2005–2015 UN International Decade for Action on 'Water for Life', it was not until July 2010 that the UN General Assembly finally adopted the resolution that 'recognized the right to safe and clean drinking water and sanitation as a human right that is essential for the full enjoyment of life and all human rights' (A/RES/64/292 of 28 July 2010). Shortly thereafter, in September 2010, the UN Human Rights Council further confirmed that it was legally binding upon states to respect, protect, and fulfill the right (A/HRC/15/L.14 of 24 September 2010). These major international policy shifts have been heralded by most people as a move in the right direction towards addressing global water inequities.

1

However, in recent years, some scholars and activists have also sounded a note of caution, bringing attention to the challenges in materializing this right, as well as questioning what it will really mean for the politics of water governance, equity and justice (see chapter by Bakker[1] in this book for a helpful summary; Anand, 2007; Bakker, 2010; Goldman, 2007; Zetland, 2010). Some build on a longer tradition of left critique of the notion of rights (Brown, 1997).[2] These are seen as inherently individualizing and, in the case of human rights, they are seen to neglect the economic injustices that permit the continued violation of people's basic dignity, building instead on a liberal democratic framework that fails to recognize the reproduction of unequal power relations within capitalist societies. In spite of these limitations, given the moral weight behind calls for the right to water, few would argue, unequivocally, against it: perhaps few would dare. Nevertheless, in what appears to be an emerging consensus around the right to water, much of the critical power within the current movement is being negated. The right to water risks becoming an empty signifier used by both political progressives and conservatives who are brought together within a shallow post-political consensus that actually does little to effect real change in water governance. This is not helped by the conflation of quite different terms when the right to water is collapsed into broader discussions of ownership of 'water rights' and more ecocentric conceptions of 'the rights of water'.[3] Responding to both concerns and critiques of the movement for the right to water as well as critiques of contemporary water governance, this book is an intervention at a crucial moment into the shape and future direction of struggles to achieve water justice.

Whilst many see the rights discourse as addressing broader issues of justice, others warn it can subvert water equity if efficiency and full-cost recovery are prioritized (PSIRU, 2002; Branco and Henriques, 2010; Spronk, 2010). Since the Dublin Principles of 1992 that, in part, framed water as an economic good, concerns have been raised that full cost recovery will further exclude the poorest from water provision. Commercialization, privatization and commodification of water has resulted in a situation where those who can pay for water have it readily, leaving many without affordable or accessible water sources. The bulk of such critiques have focused on the effects of privatization of municipal utilities, the growth of the bottled water industry, and the trading of water as a commodity, all of which have contributed to the calls for water to be held in the commons and as a public trust (for greater detail, see Barlow and Clarke, 2002; Shiva, 2002; Barlow, 2008). Polarizing pro- and anti-privatization debates, often framed in terms of commodification-versus-rights, have ensued in academic and policy circles in recent years. Critical attention was brought to how and why certain modalities are followed and with what outcomes vis-à-vis financing water provision as well as the impacts on the lives of vulnerable groups (Bond and Dugard, 2008; Hall and Lobina, 2006). Many continue to see the rights discourse as necessarily addressing broader issues of justice, while being critically watchful of the capture of rights discourses by powerful for-profit market forces in implementation plans or policy designs. As the disabling dualisms of the public-versus-private debate continue to polarize many interventions (for criticisms of such dualisms, see Budds and McGranahan, 2003; Swyngedouw, 2007; Bakker, 2010), some scholars have focused their attention on the reinvigoration and reclaiming of public stewardship (e.g. Balanyá et al, 2005) while others are investigating alternatives to privatization that does not necessarily mean going back to the older forms of public provisioning (e.g. McDonald and Ruiters, 2011). Overall, concerns continue to exist over the role of the market, private sector and for-profit provision of water vis-à-vis commodification processes that could co-opt the right to water, whereby commercialization and privatization of water ends up coming in the wake of making water a right, thereby subverting goals of water justice.

Ever since the emergence of calls for the right to water, critics have in the above-mentioned ways shown how some of the demands can obfuscate as much as they clarify, perhaps furthering the very agendas that water justice activists seek to counter. In this regard, some have made the point that major corporate interests are among the more unlikely – and yet most vocal – supporters of the right to water as a means for greater expansion of business opportunities: in this case a struggle to achieve fair access to water is in danger of producing its own nemesis (Morgan, 2004; Mehta, 2005; Bond and Dugard, 2008; Russell, 2011). Thus, when in 2010, Catarina de Albuquerque (the UN's Independent Expert on the issue of human rights obligations related to access to safe drinking water and sanitation) stated that there were no prescriptive models of service provision (A/HRC/15/31 of 29 June 2010), concerns emerged whether this opened the floodgates to further commercialization within the water sector. Such concerns are real, as rights discourses do not necessarily preclude marketization, privatization or dispossession. This, in turn, as we demonstrate later, underscores the need to rearticulate debates with political questions around democracy, justice and equity.

It would be naïve in this context to assume that private sector participation and the influence of for-profit water industries will be negated by achieving legal recognition of the right to water: indeed the response of the global water industry to the UN's resolution is somewhat disconcerting. Immediately following the 2010 UN resolution, *Global Water Intelligence*, a magazine that promotes private water investment, took the opportunity to reassure investors that it represented a 'massive defeat for the Global Water Justice Movement' (Global Water Intelligence, 2010a; Global Water Intelligence, 2010b).[4] The reasoning behind this: the right to water remained fundamentally compatible with private sector participation and contained no obligation on utilities to provide subsidies to poor communities. Therefore, if rights frameworks can outline the basic issues and provide legitimacy to pursuing equitable water allocation, they do not guarantee that there will be fair implementation or that co-optation by powerful forces will be prevented from subverting water justice goals (Morgan, 2004; Gupta et al, 2010). Indeed the right to water says little about how people might be provided with water and who will provide this (Dubreuil, 2006). While learning from 'good practices' can become part of a new dialogue, it becomes imperative to be alert to problematic implementation plans or policies. Without imputing such critical meaning, even in contexts in which the right to water has been recognized by national governments and the international community, the achievement of this has the potential to fail to bring the hoped-for radical transformation of equitable access to safe water (Mehta and Madsen, 2005; Winkler, 2008). Simultaneously, it is vital to question the conflation with polyvalent and contentious notions of development, participation, community, empowerment and sustainability, since water policies often invoke such terms (Molle, 2008; Sultana, 2009; Clark's chapter in this book). While such notions can enable the discursive thrusts to push for more equitable water provisioning, a critical eye has to be maintained on what these translate to on the ground and how they are reified or critiqued in any given context in a globalizing world. A reflective praxis in materializing the right to water thus becomes essential. This is a central aim of our book.

The move towards making the right to water legally binding means that concrete action on the policy imperatives becomes important for institutions and nation-states; however, it also highlights the challenges inherent in operationalizing the universal call for a right to water. While the right to water is often deemed anthropocentric and contentious, the discursive and policy spaces created through such debates enable more equitable possibilities to be struggled for, envisioned, and plausible tactics for distributive justice and democratic processes to be pursued. Nonetheless, the legal instruments, institutions, processes and outcomes need to be critically and carefully analyzed contextually (Langford, 2005; Ingram et al, 2008). Factors such as availability, accessibility, acceptability, appropriateness, affordability and quality are often highlighted in policy overtures as being inherent in discussions over the right to water (e.g. COHRE, 2007), but these cannot be assumed or taken for granted, rather they have to be negotiated and realized in any given context (e.g. Bell et al, 2009). As a result, raising incisive questions of process, mechanism, actors, scale, exclusions and politics that are imbricated in struggles over water thus come to the forefront in any materialization or reconfiguration of the right to water. This in turn highlights the importance of law, legal systems, property relations and governance structures (e.g. see chapters in this book by Bakker, Mitchell, Schmidt, Linton, Staddon et al, and van Rijswick and Keessen). The debates around the right to water in general underscore the need for greater focus on power relations in decision-making about water, who gets water and who does not, how water becomes accessible or available, with what means and ends, and how water governance is enacted across sites and scales. Recognizing the right to water signals that authorities can be held politically and legally accountable, enabling those who are denied water to have means to contest and struggle for water. Opportunities can be created for marginalized communities and peoples to enter into (often elitist) decision-making processes of water policies, management systems and institutions. Most scholars and activists point out that the spirit of the debates around the right to water are to highlight that pro-poor and equitable water access be ensured, whereby multiple actors and processes can converge to rearticulate the specificities of a context, but embody the general concerns of equality, social justice and deep democracy (cf. Appadurai, 2001).

Justice, politics and struggles

Within this context, global struggles over water have, however, taken different forms, reworking spaces, scales and peoples in complex ways, underscoring that discursive and material struggles over water are bound up with questions of power and governance. In this regard, a scalar politics has emerged in which struggles actively produce new forms of water governance. While struggles for the right to water can articulate with specific historical geographies, they simultaneously connect with broader global concerns and universal rights discourses. While holding governments legally accountable is made possible in the recent global resolutions, these are often only actualized through social struggles that translate moral arguments over rights to water into workable claims. In turn, new relationships are forged between citizens and states, and a range of actors (such as non-governmental entities and grassroots organizations) have increasingly entered into the debate (cf. Keck and Sikkink, 1998). A global water justice movement has emerged from such concerns and critiques.

Defining the global water justice movement, Barlow (2008, pp xi–xii) states that the movement consists of 'environmentalists, human rights activists, indigenous and women's groups, small farmers, peasants and thousands of grassroots communities fighting for control of their local water sources. Members of this movement believe that water is the common heritage of all humans and other species, as well as a public trust that must not be appropriated for personal profit or denied to anyone because of inability to pay.' Such calls emerge from the massive inequities in water provision and access, where high water prices in for-profit provision systems have led to water-related marginalization, suffering and death. Calls for greater public reinvestment, accountability, transparency, monitoring and regulation are often built into goals of the water justice movement, as well as an implicit recognition of the value and sanctity of water for both society and nature (see also Shiva, 2005). Barlow (2008) points out that critical attention is needed on concerns of displacement, mismanagement and capture of water, with continued attention to issues of power and control: who has it, who does not, who benefits or loses, in what ways, and to what effect. This becomes constitutive of the re-evaluations of the priorities, visions, and principles that guide water governance in any context. In the goals of democratizing water regulation, management and policy-making, a reflexive practice thus becomes imperative. Such underscoring of the need to deconstruct given systems and engender critical debate are important to the water justice movement. To this end, the UN's recognition of the right to water is viewed as a moral statement in recognizing the importance of prioritizing water for life, and as a way to foster transforming the dominant way water has been viewed as a commodity and challenging its valuation as a purely economic good. In addition, the role of the state and other actors involved in water policy-making, management and provision, especially to marginalized and vulnerable groups, are brought to the fore in reconfiguring equitable allocation, access and use of safe water. Beyond this, holding water in the public trust, with a not-for-profit governance system, are often articulated by advocates of the water justice movement (for example, 'Take back the tap' projects that call attention to reinvesting in public infrastructure and good governance in explicit critiques of the bottled water industry; see Food and Water Watch, 2009; Bell et al, 2009). Thus, the dual roles of critique and advocacy are entwined.

We are sympathetic to such epistemological and political concerns. However, we begin with an acute sense of the dangers of terminological slippage, of the banalities of some claims to the right to water and of the dangers of deliberate or naïve political misappropriation of the water justice movement's gains. Nevertheless, rather than rejecting struggles for the right to water, the difficulties and ambiguities are seen as the starting point for developing a more sound political footing. Our general stance is characterized by a cautious optimism: a new movement is emerging but this is one that has many challenges yet to confront. In this regard, the chapters in this book are bold, provocative and yet contemplative. Rather than reactive to the efforts to co-opt the struggle for the right to water, the book aims to be pro-active in defining what this struggle *could* mean and how it might be taken forward in a far broader transformative politics. Above all, within this, we question the immanent potentials in local, national and global struggles for the right to water, thereby enhancing understanding and insights on the ways in which a global movement is influenced and shaped by local political, economic and cultural dynamics. We seek to elucidate how universal calls for rights articulate with local historical geographical contexts, and the barriers and potentials that emerge from this. In recognizing the importance that water activists place on the concept of rights, we seek to engage productively with, rather than dismissing, the human right to water. Many argue that the question of rights has become a terrain for debate and political contestation and, therefore, potentially, a platform for democratizing water debates. Rather than foreclosing possibilities, this book is replete with critical opportunities. As Harvey (2000) has noted, the maelstrom of contradictions opened up by the question of rights can serve as a prelude to a far more radical, transformative political project. In short, our aim is to bring a geographical sensitivity to calls for a universal right to water: within this, we see the right to water as one necessary but insufficient moment in the struggle to achieve equitable access to water for all.

We take such an approach forward through a range of chapters that focus on philosophical framings (chapters by Bakker, Schmidt, Linton), the role of law and legal frameworks (chapters by Staddon et al, van Rijswick and Keessen, Ruru) and the question of property relations and civil society (chapter by Mitchell), before integrating some of these more abstract arguments with a range of concrete struggles (chapters by Giglioli, Meehan, Clark, Bond, Bywater, Perera, Bustamante et al). The early chapters engage with a range of epistemological positions. Here, the theoretical paradoxes and pitfalls are considered and a debate is opened up over the direction of future demands, with a review of how such foundations have been captured within new forms of water governance. We then move to work through such perspectives empirically. Here, a range of studies are mobilized that integrate more abstract questions to the realities of everyday life, grounding the theoretical debates in order to enrich current conceptualizations and discourses. Through the empirical examples from Africa, Asia, Oceania, Latin America, the Middle East, North America and the European Union, we argue that calls for a human right to water in differing geographical contexts can inform broader political endeavors, thereby demonstrating the increased geographical sensitivity to calls for a universal right to water. In each of these contexts, activists and policy-makers have sought to define, through processes of negotiation and contestation, what is meant by the right to water. Transforming the 'right to water' from an empty signifier to a powerful tool for mobilizing from the grassroots, such struggles have gone well beyond the new rights-based approaches to development (e.g. see chapters by Bustamante et al, Bywater, Giglioli, Meehan). Indeed, they can be seen as at the cutting edge of a new networked politics crossing geographical locations and narrow disciplinary concerns (e.g. see chapter by Perera) or different ways of relating to water (e.g. see chapter by Ruru). Often building on the paradoxes that are opened up within rights-based discourses, scholars and activists have sought to give real meaning to the right to water whilst broadening what is seen as a democratic core in the movement for water justice. In the South African example, for instance, ever since the country's new constitution was scripted in 1996 activists have sought to use 'the right to water' as a means of defining a new direction for the ANC government's post-apartheid policy-making (e.g. see chapters by Clark and Bond; Loftus and Lumsden, 2008).

All of these cases further enrich and contribute to existing framings in our understandings of the right to water. Throughout, all contributors seek to reclaim the ground on which the right to water will be defined in coming years, applying their critical tools in order to wrest it away from a narrowly defined, technocratic realm. In concluding this introduction, we suggest several areas around which future debates might find some common ground. These build on the following points. First, there seems to be a crucial desire to ensure that the cry for the right to water does not descend into meaningless technical discussions that deaden the transformative potentials within the emerging movement. In many respects, this brings us squarely into questions of what constitutes the truly political. If the call for the right to water is to become a genuinely political moment, we need to consider how it might acquire a material force within the world and how it might become actually world-changing. Secondly, and this is perhaps implicit throughout what has been said, we need to consider ways in which specific struggles for the right to water work with, are shaped by, and influence global struggles for this right. Thirdly, if we succeed in reclaiming the right to water from the technocratic realm to which it is in danger of being consigned, and if we ensure it makes that move from the local to the universal without shunning questions of difference, then the right to water has the potential to mean far more than achieving access to sufficient volumes of safe water. Potentially, it means the right to be able to participate more democratically in the making of what Linton (2010 and in this book), amongst others, terms the 'hydrosocial cycle'[5] (see also Swyngedouw, 2004). The right to water could mean the right to transform the socionatural conditions out of which water is currently accessed. In this sense, it means a remaking of our relations with human and non-human others. In short, it might assume a role in the remaking of our world in more fair, just and democratic ways. In this regard, we remain hopeful that existing scholarship on water governance and water struggles will fruitfully inform further research, activism, and the making of more egalitarian and just water futures. Interdisciplinary critical scholarship on water is both broad and deep, and substantive insights can be drawn from such bodies of writing to inform debates on the right to water (for instance, Gandy, 2002; Mosse, 2003; Strang, 2004; Swyngedouw, 2004; Conca, 2005; Kaika, 2005; Castro, 2006; Baviskar, 2007; Bakker, 2010; Linton, 2010; Johnston, 2011). While these interventions may not directly articulate with debates around the right to water, they provide insights that can enrich current conceptualizations.

In thinking through the challenges of materializing a right to water, attention to the intersectionalities with multiple processes and forces can critically elucidate possible ways forward (cf. Salzman, 2006; D'Souza, 2008; Derman and Hellum, 2008). For instance, the ways that the right to water coalesces around, intersects with and transforms or challenges other rights (e.g. gender rights) are important signifiers in the ongoing struggles over the right to water (Brown, 2010). The impacts of water insecurity and injustices are clearly gendered, where women and girls in much of the global South

8

spend countless hours fetching water for productive and reproductive needs. A gendered division of labor, as well as gendered livelihoods, wellbeing and burdens, are deeply affected by water quality, availability, provision systems and water policies (Crow and Sultana, 2002; O'Reilly et al, 2009; Cleaver and Hamada, 2010; Sultana, 2011). Gender intersects with other axes of social difference (such as class, race, caste, dis/ability, etc.) whereby water crises can exacerbate socially constructed differences and power relations. Similarly, social struggles over the right to water are gendered, articulating with contextual social differences that shape the nature and outcomes of struggles (Laurie, 2011). Historically and geographically situated practices that are defined in relation to water (from the politics of mega-dams to the practice and politics of collecting water) influence everyday life in complex ways. Scholars have therefore argued that multiple, situated and place-based struggles thus can link and contribute to transnational movements (cf. Mohanty, 2003; Harcourt and Escobar, 2005), where difference and diversity are constitutive of the broader calls of equality in the right to water.

Throughout the book we make explicit the *conjunctural* nature of struggles for the right to water. Struggles articulate with a set of local and regional discourses around the value of water and the meaning of individual and collective rights within each of the contexts. In this regard, the geographical specificities come to the forefront of each chapter whilst they also explore some of the subtle and nuanced scalar politics at play in bringing together militant particularist (cf. Harvey, 1996) demands with global ambitions for fairer and equitable allocation, access and management of water. Within activist positions, again, the complex political positioning needed is dwelt upon and explored. In this context, it is interesting to note how the right to water 'travels', with the South African example being used as both an inspiration and a salutary lesson in different contexts. Wary of Said's (1983) cautions around the loss of critical edge in 'travelling theory', each of the chapters seeks to better understand the complex geographical imaginations and the particular articulations when rights-based discourses travel.

The right to water: floating signifier, bureaucratic rationality or political possibility?

As we have alluded to, most people would agree that the right to water is, in principle, a good thing; however, the concept seems to mean quite different things at different times and in different places (Naidoo, 2010). Thus, the key challenge is to be able to fill this empty signifier with real political content. Such content must build on the historically and geographically specific practices of those currently seeking to achieve fair access to water and, if water justice activists are to define it, this will involve reclaiming 'the right to water' from the technocrats who are currently seeking to script it. Instead, activists need to ensure struggles for the right to water are shaped by the efforts of those for whom it offers freedom from the nightmares of their history. Here, we might think of the veterans of the Cochabamba Water Wars described in the chapter by Bustamante et al, or the cosmopolitan subalterns described in the chapters by Perera or Bywater. Indeed, the book charts many such movements: here, we begin to witness the constitutive role of subaltern struggles for indigenous rights to water (see the chapter by Ruru) or efforts to reshape broader geopolitical configurations (see the chapter by Giglioli) and also to challenge the criminalization of efforts to subvert the state hydraulic paradigm (see the chapter by Meehan). Perhaps most starkly, the South African examples show the dangerous ambiguities remaining if we leave this signifier floating. As both Clark and Bond show in different ways in their chapters, the constitutional guarantee of the right to water in South Africa remains hollow for many of the residents of informal settlements and townships where new forms of violence (ranging from the perversely titled self-disconnection to the aggressive installation of flow-limiting devices) have accompanied the victory of activists in securing their rights (see also Loftus, 2006).

In seminal contributions to these debates, Bakker (2010 and chapter in this book), elaborates on the pitfalls in a growing movement for the right to water. Perhaps the key point Bakker makes is similar to Naidoo (2010): the right to water has such a shifting meaning that it allows for agreement between anyone, from large multinational water companies seeking to bid for concession contracts in cities of the global South to activists within those cities fighting the privatization of their municipal services. We are all for the right to water – from the vendor selling from his tanker to the thirsty activist seeking radical change. Lacking specificity, the right to water loses its conceptual weight: it becomes a floating signifier devoid of any political content. Like 'sustainable development' and many other fuzzy concepts that have gone before, the right to water is emptied of any real meaning. If all concur it is a good thing it loses its ability to disrupt contemporary water governance which has persistently reproduced inequities.

As detailed in the chapter by Bustamante et al, this debilitating consensus implies a post-political moment. Working with the conceptual tools that have emerged in recent post-marxist debates, as well as the grounded realities of activists' disappointments with the Bolivian government's continuing concessions to mineral extraction industries, these authors add much to the ground already staked out by scholars such as Bakker. Turning to Rancière (2004), they demonstrate that the truly political would involve the disruption of the 'police' distribution of the sensible. This implies a dissensual politics, differing radically from one operating within the given police order and shifting from a politics of demands, directed at and to be granted by the given order, to one that actively seeks to transform this order. Rancière is not the only political thinker to be engaged in such discussions and these debates have been taken forward incredibly effectively within both geographical and environmental thought. Swyngedouw (2010), for example, argues that discussion of climate change is essentially post-political. Most of the positions taken over climate change or, even more so, of environmental sustainability have virtually no concern for transforming that which is given to the sensible. Rather, 'debate' involves reconciling oneself to the given order of things whilst operating on an increasingly limited terrain: the question is not how to achieve a radically different world but rather how to make sure the current world is reproduced in low-carbon ways. In many ways the climate justice movement parallels the water justice movement, and there are points of overlap, and potential pitfalls, shared by the two.

The key challenge in this respect is to ensure that the right to water comes to refer to a genuinely political activity, one through which we might rethink the very foundations on which the world is sensed, made sense of and lived. Again, as pointed out by many others (e.g. Naidoo, 2010; Barlow, 2011), activists should work to put this appropriation of political content at the forefront of struggles for the right to water. Nevertheless, if we are to limit ourselves to recent philosophical attempts to demarcate the genuinely political, there is a grave risk that we might overemphasize ruptures and dissonance over actually existing practices. Thus, we could find ourselves developing an intellectually refined position that is actually at odds with the views of those working at the grassroots. In some respects, this is exactly the point made by Rancière: 'A dissensus is not a conflict of interests, opinions or values; it is a division put in the "common sense": a dispute about what is given, about the frame within which we see something as given' (2004, p 304). Indeed one of the more disabling moves of parliamentary democracy has been to reduce politics to a polite exchange of differing views, bringing the grassroots perspective onboard, under a façade of genuine equality. But perhaps such a 'division', as Rancière puts it, risks essentializing 'the political' and divorcing conceptual critique from the movement on which the future of the right to water will surely rest.

The work of Antonio Gramsci may be insightful here. As with Rancière (1989) in *The Nights of Labour*, and certainly with the contributors in this book, our starting point must be those empirical realities. Rather than beginning with a division put in the common sense, Gramsci's intention is to build a transformative politics from within the shards of existing common sense. Here, he begins from the always contradictory realities and ways of thinking that exist on the ground and builds a 'philosophy of praxis' whose aim is to bring coherence (in this case meaning the identity of theory and practice) to the incoherent realm of common sense.[6] Good sense emerges from within common sense in an immanent critique that moves dialectically between theory and empirical reality. More than ever, we need this form of dialectical pedagogy within struggles for the right to water. In searching for a radically new conception of the right to water, we may well think of Gramsci's observation:

11

Is it possible that a 'formally' new conception can present itself in a guise other than the crude, unsophisticated version of the populace? And yet the historian, with the benefit of all necessary perspective, manages to establish and to understand the fact that the beginnings of a new world, rough and jagged though they always are, are better than the passing away of the world in its death throes and the swan-song that it produces.

(Gramsci, 1971, pp 342–3 Q11 §12)[7]

Gramsci's thought is animated by a germinating historical geographical materialism. The same sensitivities to both history and geography are also needed if the right to water is to be able to achieve a politics that works on a global stage without eliding the very differences and specificities that have animated struggles in radically different contexts. In this respect, we need to consider how 'militant particularisms' might be effectively translated into global ambitions. In many respects, one of the most inspiring aspects of the call for the right to water has been its ability to move across, whilst also disrupting, a scalar politics. If Bakker (2010) is partly right in suggesting that struggles for water justice have been somewhat less effective than anti-dams campaigns, appearing less networked and only weakly articulated within global campaigns, this must also be viewed alongside the remarkable mobilization that resulted in the UN resolution in July 2010. Here, a broad coalition of geographically disparate activists managed to coalesce in a struggle to make a truly international politics. Of course the dangers, as Bakker also points out, are that locally specific practices of governing water might be lost if this universal call is actually effective in achieving change at the grassroots level. A world of contradictions is opened up and we need to think carefully about how to navigate this. Gramsci's attempts to think through these questions draw heavily from the politics of conjunctures that animates Marx's most incisive political commentary, *The Eighteenth Brumaire of Louis Bonaparte* (Marx, 1974). Here, Marx follows the temporal rhythms through which a working class politics developed in France between 1848 and 1851. Both long-term and short-term processes work together to shape the limited potentials for revolutionary change. Gramsci takes this forward in his reading of the Risorgimento in Italy and the reversal of the revolutionary moment that decided his own fate in the 1920s. However, for Gramsci, neither time nor space are prioritized, rather they are internally related. Political struggles wax and wane not simply over time but through their relations with other movements and other ideas operating in, and ultimately producing, both time and space. This seems absolutely crucial for a politics of the right to water that might have some conceptual weight in a range of different contexts. We should be clear here: we are not arguing for one second that Gramsci, or Rancière for that matter, has any of the answers for thinking through the most pressing questions facing activists and communities struggling for the right to water. But in a philosophy of praxis that is able to articulate a range of historically and geographically specific subaltern practices and conceptions, we do find fertile suggestions for thinking through the ways of working through these questions ourselves – as a co-conspiratorial group of activists and academics.

Beyond this, Gramsci also suggests a politics in which the non-human and human are inseparable (Fontana, 1996; Ekers et al, 2009): his is a politics of socionatures, not one in which the social is divorced from the human in some impossible-to-sustain antinomian framework. The same cannot always be said of Henri Lefebvre's (1991) politics, whose scholarship, as Bond's chapter demonstrates, does nevertheless resonate in different ways with the debates around the right to water. Yet again, Lefebvre's writings have seen renewed interest in recent years: this time because of his passionate cry for the right to the city (Lefebvre, 1996) which has, as with the right to water, galvanized activists from the favelas of Rio to the campuses of Manhattan. Bond's chapter suggests some of the potential common grounds between the right to the city movement and the calls for the right to water. Above all, what seems to animate Lefebvre's reconfigured notion of the right to the city, is not a narrowly conceived notion of the right of people to reside in cities. Rather it is the right of all to be able to participate in the making of cities, conceived as oeuvres rather than static entities (see also Harvey, 2008). In the process, this new urban life comes to be reflected in the subjectivities of those actively participating in its making. Here, we see a model of mutual co-production between urban form and urban dweller. Reconfiguring the right to water on the same grounds, whilst also recognizing (as Lefebvre did not (see Smith, 1997)) that this is a fundamentally socionatural, as opposed to purely social, activity would then provide a radical base from which to work towards the articulation of radically distinct subaltern perspectives in the democratization of the hydrosocial cycle.

It might seem a long shot to suggest that the right to water holds out the hope of remaking our world. But for many water justice activists, this is what makes the movement a truly political one. We have an obligation to build on such struggles rather than simply using them for our own intellectual debates. To this end, our hope is that the book contributes to and continues the journey of intellectual and political projects that think through and materialize this right to water: understood as a political moment, akin to the right to the city, and implying democratic participation in producing the flows of water and social power on which life itself depends.

Notes

1 Bakker's chapter is a reproduction of her seminal 2007 article, with a new postscript at the end. The rationale behind reproducing the text is so that readers have easy access to this 'artifact' that had sparked considerable debate and interest. The

postscript added by Bakker in the chapter reflects the evolution of her thoughts since 2007.

2 For Marx's criticisms – in a profoundly different moment – see *The Critique of the Gotha Programme* (Marx, 1974); and for deliberately overdrawn caricatures of various positions, see Lukes, 1997.

3 The distinction between 'right to water' and 'water rights' are important to note, as the former focuses more on issues of human rights, access to safe drinking water, equity and justice, whereas the latter often has an economistic/legalistic focus on contractual obligations, concessions, property rights and water markets. While these distinctions are often blurry, and both are wrapped up with water struggles, we believe it is important to recognize the differences in terminologies and tropes.

4 We are grateful to Cristy Clark for this input.

5 In Linton's (2010, p 68) terms, this 'describes the process by which flows of water reflect human affairs and human affairs are enlivened by water.'

6 For feminist elaborations on the notion of praxis, see Nagar et al (2002); Mohanty (2003); Harcourt and Escobar (2005).

7 In line with recent Gramsci scholarship, this reference includes both the most readily available source in English for the notebooks and also the notebook and note number from the critical edition (not yet available in English).

References

Anand, P. (2007) 'Right to water and access to water: an assessment', *Journal of International Development*, vol 19, pp 511–526.

Appadurai, A. (2001) 'Deep democracy: urban governmentality and the horizon of politics', *Environment and Urbanization*, vol 13, no 2, pp 23–43.

Bakker, K. (2007) 'The "commons" versus the "commodity": alter-globalization, anti-privatization and the human right to water in the global South', *Antipode*, vol 39, no 3, pp 430–455.

Bakker, K. (2010) *Privatizing Water: Governance Failure and the World's Urban Water Crisis*, Cornell University Press, Ithaca, NY.

Balanyá, B., Brennan, B., Hoedeman, O., Kishimoto, S. and Terhorst, P. (2005) *Reclaiming Public Water: Achievements, Struggles and Visions from around the World*, Transnational Institute and Corporate Europe Observatory.

Barlow, M. (2008) *Blue Covenant: The Global Water Crisis and the Coming Battle for the Right to Water*, The New Press, NY.

Barlow, M. (2011) *Our Right to Water: A Peoples' Guide to Implementing the United Nations Resolutions on the Right to Water and Sanitation*, Council of Canadians, Ottawa.

Barlow, M. and Clarke, T. (2002) *Blue Gold: The Battle Against the Corporate Theft of the World's Water*, Earthscan, London.

Baviskar, A. (2007) *Waterscapes: The Cultural Politics of a Natural Resource*, Permanent Black, New Delhi.

Bell, B., Conant, J., Olivera, M., Pinkstaff, C. and Terhorst, P. (2009) *Changing the Flow: Water Movements in Latin America*, Food and Water Watch, Other Worlds, Reclaiming Public Water, Red VIDA, and Transnational Institute.

Bond, P. (2008) 'Macrodynamics of globalisation, uneven urban development and the commodification of water', *Law, Social Justice & Global Development Journal* (LGD), www.go.warwick.ac.uk/elj/lgd/ 2008_1/bond, accessed 20 December 2009.

Bond, P. and Dugard, J. (2008) 'Water, human rights and social conflict: South African experiences', *Law, Social Justice & Global Development Journal* (LGD), www.go.warwick.ac.uk/elj/lgd/2008_1/bond_dugard, accessed 20 December 2009.

Branco, M. and Henriques, D. (2010) 'The political economy of the human right to water', *Review of Radical Political Economics*, vol 42, pp 142–155.

Brown, C. (1997) 'Universal human rights: a critique', *The International Journal of Human Rights*, vol 1, no 2, pp 41–65.

Brown, R. (2010) 'Unequal burden: water privatization and women's human rights in Tanzania', *Gender and Development*, vol 18, no 1, pp 59–67.

Budds, J. and McGranahan, G. (2003) 'Are the debates on water privatization missing the point? Experiences from Africa, Asia and Latin America', *Environment and Urbanization*, vol 15, no 2, pp 87–113.

Castro, J. (2006) *Water, Power and Citizenship: Social Struggle in the Basin of Mexico*, Palgrave Macmillan, Basingstoke.

Cleaver, F. and Hamada, K. (2010) '"Good" water governance and gender equity: a troubled relationship', *Gender and Development*, vol 18, no 1, pp 27–41.

COHRE (2007) *Manual on the Right to Water and Sanitation*, Publication of the COHRE, AAAS, SDC and UN-HABITAT, www.cohre.org, accessed 20 December 2009.

Conca, K. (2005) *Governing Water: Contentious Transnational Politics and Global Institution Building*, MIT Press, Cambridge.

Crow, B. and Sultana, F. (2002) 'Gender, class and access to water: three cases in a poor and crowded delta', *Society and Natural Resources*, vol 15, no 8, pp 709–724.

Derman, B. and Hellum, A. (2008) 'Observations on the intersections of human rights and local practice: a livelihood perspective on water', *Law, Social Justice & Global Development Journal* (LGD), www.go.warwick.ac.uk/elj/lgd/2008_1/derman_hellum, accessed 20 December 2009.

D'Souza, R. (2008) 'Liberal theory, human rights and water-justice: back to square one?', *Law, Social Justice & Global Development Journal* (LGD), www.go.warwick.ac.uk/elj/lgd/2008_1/desouza, accessed 20 December 2009.

Dubreuil, C. (2006) *The Right to Water: From Concept to Implementation*, World Water Council, France.

Ekers, M., Loftus, A. and Mann, G. (2009) 'Gramsci lives!', *Geoforum*, vol 40, pp 287–291.

Fontana, B. (1996) 'The concept of nature in Gramsci', *Philosophical Forum*, vol 27, pp 220–243.

Food and Water Watch (2009) *Dried Up, Sold Out: How the World Bank's Push for Private Water Harms the Poor*, Food and Water Watch, Washington DC.

Gandy, M. (2002) *Concrete and Clay*, MIT Press, Cambridge.

Gleick, P. (1999) 'The human right to water', *Water Policy*, vol 1, pp 487–503.

Global Water Intelligence (2010a) 'The human right to a national water plan', www.globalwaterintel.com/insight/human-right-national-water-plan.html, accessed 20 August 2010.

Global Water Intelligence (2010b) 'Another bad idea we need to act on', www.globalwaterintel.com/insight/another-bad-idea-which-we-need-act.html, accessed 20 August 2010.

Goldman, M. (2007) 'How "water for all!" policy became hegemonic: the power of the World Bank and its transnational policy networks', *Geoforum*, vol 38, pp 786–800.

Gramsci, A. (1971) *Selections from the Prison Notebooks of Antonio Gramsci*, Lawrence and Wishart, London.

Gupta, J., Ahlers, R. and Ahmed, L. (2010) 'The human right to water: moving towards consensus in a fragmented world', *Review of European Community & International Environmental Law*, vol 19, no 3, pp 294–305.

Hall, D. and Lobina, E. (2006) *Pipe Dreams: The Failure of the Private Sector to Invest in Water Services in Developing Countries*, Public Services International Research Unit, World Development Movement, London.

Harcourt, W. and Escobar, E. (2005) *Women and the Politics of Place*, Kumarian Press, CT.

Harvey, D. (1996) *Justice, Nature and the Geography of Difference*, Blackwell, Oxford.

Harvey, D. (2000) 'Uneven geographical developments and universal rights', in *Spaces of Hope*, University of California Press, Berkeley.

Harvey, D. (2008) 'The right to the city', *New Left Review*, vol 53, pp 23–40.

Ingram, H., Whiteley, J. and Perry, R. (2008) 'The importance of equity and the limits of efficiency in water resources', in J. M. Whiteley, H. Ingram and R. Perry (eds) *Water, Place, and Equity*, MIT Press, Cambridge.

Johnston, B. (2011) *Water, Cultural Diversity & Global Environmental Change: Emerging Trends, Sustainable Futures?*, Springer, Netherlands.

Kaika, M. (2005) *City of Flows*, Routledge, London.

Keck, M. and Sikkink, M. (1998) *Activists without Borders: Advocacy Networks in International Politics*, Cornell University Press, Ithaca, NY.

Langford, M. (2005) 'The United Nations concept of water as a human right: a new paradigm for old problems?', *International Journal of Water Resources Development*, vol 21, no 2, pp 273–282.

Laurie, N. (2011) 'Gender water networks: femininity and masculinity in water politics in Bolivia', *International Journal of Urban and Regional Research*, vol 35, no 1, pp 172–188.

Lefebvre, H. (1991) *The Production of Space*, Blackwell, Oxford.

Lefebvre, H. (1996) *Writings on Cities*, translated and edited by E. Kofman and E. Lebas, Blackwell, Oxford.

Linton, J. (2010) *What is Water? The History of a Modern Abstraction*, University of British Columbia Press, Vancouver.

Loftus, A. (2006) 'Reification and the dictatorship of the water meter', *Antipode*, vol 38, pp 1023–1045.

Loftus, A. and Lumsden, F. (2008) 'Reworking hegemony in the urban waterscape', *Transactions of the Institute of British Geographers*, vol 33, no 1, pp 109–126.

Lukes, S. (1997) 'Five fables on human rights', in M. Ishay (ed.) *The Human Rights Reader*, Routledge, London.

McDonald, D. and Ruiters, G. (2011) *Alternatives to Privatization in the Global South*, Routledge, New York.

Marx, K. (1973) *Surveys from Exile: Political Writings Volume 2*, Pelican, London.

Marx, K. (1974) *The First International and After: Political Writings Volume 3*, Pelican, London.

Mehta, L. (2005) 'Unpacking rights and wrongs: do human rights make a difference? The case of water rights in India and South Africa', *IDS Working Paper 260*, Institute of Development Studies, Brighton.

Mehta, L. and Madsen, B. (2005) 'Is the WTO after your water? The General Agreement on Trade in Services (GATS) and poor people's right to water', *Natural Resources Forum*, vol 29, no 2, pp 154–164.

Mohanty, C. (2003) *Feminism Without Borders: Decolonizing Theory, Practicing Solidarity*, Duke University Press, Durham, NC.

Molle, F. (2008) 'Nirvana concepts, narratives and policy models: insight from the water sector', *Water Alternatives*, vol 1, no 1, pp 131–156.

Morgan, B. (2004) 'The regulatory face of the right to water', *Journal of Water Law*, vol 15, pp 179–187.

Mosse, D. (2003) *The Rule of Water: Statecraft, Ecology and Collective Action in South India*, Oxford University Press, New Delhi.

Nagar, R., Lawson, V., McDowell, L. and Hanson, S. (2002) 'Locating globalization: feminist (re)readings of the subjects and spaces of globalization', *Economic Geography*, vol 78, no 3, pp 257–284.

Naidoo, A. (2010) 'The human right to water: exploring the challenges, opportunities and implications for water justice', Plenary Lecture at *The Right to Water Conference*, Syracuse University, 29–30 March 2010.

O'Reilly, K., Halvorson, S., Sultana, F. and Laurie, N. (2009) 'Introduction: global perspectives on gender-water geographies', *Gender, Place, and Culture*, vol 16, no 4, pp 381–385.

Petrella, R. (2001) *The Water Manifesto: Arguments for a World Water Contract*, Zed Books, London.

PSIRU (2002) *Water and the Multinational Companies*, Public Services International Research Unit (PSIRU), University of Greenwich, London.

Rancière, J. (1989) *The Nights of Labour: The Worker's Dream in Nineteenth Century France*, Temple University Press, Philadelphia.

Rancière, J. (2004) 'Who is the subject of the rights of man?', *The South Atlantic Quarterly*, vol 103, no 2/3, pp 297–310.

Russell, A. (2011) 'Incorporating social rights in development: transnational corporations and the right to water', *International Journal of Law in Context*, vol 7, pp 1–30.

Said, E. (1983) 'Travelling theory', in *The Word, the Text and the Critic*, Harvard University Press, Cambridge, MA.

Salzman, D. (2006) 'Thirst: a short history of drinking water', *Yale Journal of Law & the Humanities*, vol 18, pp 94–121.

Shiva, V. (2002) *Water Wars*, South End Press, London.

Shiva, V. (2005) *Earth Democracy*, Zed Books, London.

Smith, N. (1997) 'Antinomies of space and nature in Henri Lefebvre's *The Production of Space*', in A. Light and J. Smith (eds) *Philosophy and Geography II: The Production of Public Space*, Lexington Books, Lanham, MD.

Spronk, S. (2010) 'Water and sanitation utilities in the global South: re-centering the debate on "efficiency"', *Review of Radical Political Economics*, vol 42, no 2, pp 156–174.

Strang, V. (2004) *The Meaning of Water*, Berg Publishers, Oxford.

Sultana, F. (2009) 'Community and participation in water resources management: gendering and naturing development debates from Bangladesh', *Transactions of the Institute of British Geographers*, vol 34, no 3, pp 346–363.

Sultana, F. (2011) 'Suffering *for* water, suffering *from* water: emotional geographies of resource access, control and conflict', *Geoforum*, vol 42, no 2, pp 163–172.

17

Swyngedouw, E. (2004) *Social Power and the Urbanization of Water: Flows of Power*, Oxford University Press, Oxford.

Swyngedouw, E. (2007) 'Dispossessing H_2O: the contested terrain of water privatization', in N. Heynen, J. McCarthy, S. Prudham and P. Robbins (eds) *Neoliberal Environments: False Promises and Unnatural Consequences*, Routledge, New York.

Swyngedouw, E. (2010) 'Apocalypse forever? Post-political populism and the spectre of climate change', *Theory, Culture & Society*, vol 27, no 2–3, pp 213–232.

UN Committee on Economic, Social and Cultural Rights (CESCR) (2002) 'General Comment No. 15: The right to water (Arts. 11 and 12 of the Covenant)' (20 January 2003), UN Document E/C.12/2002/11.

UN General Assembly (2010) 'Human right to water and sanitation' (28 July 2010), UN Document A/RES/64/292.

UN Human Rights Council (2010) 'Report of the Independent Expert on the issue of human rights obligations related to access to safe drinking water and sanitation, Catarina de Albuquerque' (29 June 2010), UN document A/HRC/15/31.

UN Human Rights Council (2010) 'Human rights and access to safe drinking water and sanitation' (24 September 2010), UN document A/HRC/15/L.14.

UNDP (2006) *Beyond Scarcity: Power, Poverty and the Global Water Crisis*, Human Development Report 2006, United Nations Development Programme (UNDP), New York.

WHO (2003) *Right to Water*, Health and Human Rights Publication Series Number 3, World Health Organization (WHO), Geneva.

Winkler, I. (2008) 'Judicial enforcement of the human right to water – case law from South Africa, Argentina and India', *Law, Social Justice & Global Development Journal* (LGD), www2.warwick.ac.uk/fac/soc/law/elj/lgd/2008_1/winkler/, accessed 20 December 2009.

Zetland, D. (2010) 'Water rights and human rights: the poor will not need our charity if we need their water', *Forbes Magazine*, www.forbes.com, accessed 28 December 2010.

2

COMMONS VERSUS COMMODITIES

Debating the human right to water

Karen Bakker

Prologue

On a rainy Friday in 2003, the world's Water and Environment Ministers met in Kyoto to discuss the global water crisis. While Ministers met behind closed doors, participants at the parallel public World Water Forum were presented with alarming statistics: water scarcity had been growing in many regions; and over 20% of the world's population was without access to sufficient supplies of potable water necessary for basic daily needs. In response, conference organizers had drafted an Inter-Ministerial declaration, based upon the view that the best response to increasing scarcity was the commercialization of water. International support for commercialization had been growing since the controversial Dublin Statement on Water and Sustainable Development in 1992, which advanced the principle that 'water has an economic value in all its competing uses and should be recognized as an economic good'. Supporters of the Dublin Principles argued that in light of endemic 'state failure' by governments supposedly too poor, corrupt, or inept to manage water, increased involvement of the private sector in water supply management was openly advocated by many conference participants.

Reflecting this shift in international water policy, private water companies had been invited to meet with government delegations, international financial institutions and bilateral aid agencies to develop solutions to the world's water problems. Yet activists – an international alliance of anti-dam activists, environmentalists, public sector unions, international 'bank-watcher' and 'anti-globalization' think tanks, indigenous peoples and civil society groups – protested the presence of the private water companies. These self-named 'water warriors' protested both inside and outside the Forum, critiquing the Forum co-organizers (the Global Water Partnership and the World Water Council) for their close ties to private companies and international financial institutions. Activists similarly targeted governments – particularly India and South Africa – whose record of water management and unresponsive governance was portrayed as one of the key obstacles to sustainable water management. Both governments and private companies were, activists argued, guilty of poor governance: unrepresentative, opaque and illegitimate decision-making processes (ironically, similar critiques were directed by the Forum organizers at activists).

Activists' protests culminated with the disruption of a planned highlight of the Forum – a plenary session chaired by Michel Camdessus (former head of the IMF) promoting active government support for increased private sector involvement in the water sector in the South (Winpenny, 2003). Chanting 'water is life', activists stormed the stage and demanded the withdrawal of the private sector, a return to local 'water democracy', a rejection of large dams as socio-economically and environmentally unsound and the recognition of water as a human right. Yet activists' calls fell largely on deaf ears. Southern and northern ministerial delegates reached consensus; including, controversially, support for private sector financing, new mechanisms for private sector involvement in water supply management and a conspicuous failure to refer to water as a human right.

Introduction: the triumph of market environmentalism?

The Kyoto Declaration embodies an increasingly dominant philosophy of development, variously termed 'liberal environmentalism' (Bernstein, 2001), 'green neoliberalism' (Goldman, 2005), or market environmentalism (Bakker, 2004, 2010): a mode of resource regulation which aims to deploy markets as the solution to environmental problems (Anderson and Leal, 2001). Market environmentalism offers hope of a virtuous fusion of economic growth, efficiency and environmental conservation: through establishing private property rights, employing markets as allocation mechanisms and incorporating environmental externalities through pricing, proponents of market environmentalism assert that environmental goods will be more efficiently allocated if treated as economic goods – thereby simultaneously addressing concerns over environmental degradation and inefficient use of resources.

Critical research on market environmentalism frames this paradigm as the 'neoliberalization of nature' (see Heynen et al, 2007; McCarthy and Prudham, 2004). The majority of this research focuses on the negative impacts of neoliberal reforms, including both environmental impacts and the distributional implications of the various forms of 'accumulation by dispossession' enacted by neoliberalization (Glassman, 2006), although some research also suggests that states can rationally administer environmental degradation and resource appropriation from local communities (Scott, 1998) or that environmental improvements can occur in the context of state re-regulation which accompanies privatization (Angel, 2000; Bakker, 2005).

This debate is particularly acute in the water sector. The increasing involvement of private, for-profit multinational water corporations in running networked water supply systems around the world has inspired fierce debate internationally. Proponents of market environmentalism in the water sector argue that water is an increasingly scarce resource, which must be priced at full economic and environmental cost if it is to be allocated to its highest-value uses and managed profitably by private companies whose accountability to customers and shareholders is more direct and effective than attenuated political accountability exercised by citizens via political representatives (Rogers et al, 2002; Winpenny, 1994). Opponents of market environmentalism argue that water is a non-substitutable resource essential for life and call for water supply to be recognized as a human right, which (they argue) both places an onus upon states to provide water to all and precludes private sector involvement (see, for example, Bond, 2002; Goldman, 2005; Johnston et al, 2006; Laxer and Soron, 2006; Morgan, 2004b).

Several conceptual questions underlie this debate. Is water a human right? If so, is private sector provision incompatible with the human right to water? What is the relationship between property rights regimes and privatization? And how can we best conceptualize and mobilize alternatives to neoliberalization? This paper explores these questions, documenting the different constructions of property rights adopted by pro- and anti-privatization advocates, questioning the utility of the language of 'human rights' and interrogating the accuracy of the (often unquestioned) binaries – rights/commodities, public/private, citizen/customer – deployed on both sides of the debate. The paper thus undertakes two tasks: the development of a conceptual framework for analysing market environmentalist reforms; and the application of this framework to the case of water supply.

The first part of the paper develops a typology of market environmentalist reforms in resource management, arguing that conceptual confusion frequently arises due to a lack of analytical precision about the wide range of ongoing reforms that are often over-simplified into a monolithic (and inaccurately labelled) 'neoliberalism'. The second section examines one example of these conceptual confusions: the positioning of 'human rights' as an antonym to 'commodities' by anti-privatization campaigners. After documenting the tactical failures of such an approach, the paper contrasts 'anti-privatization' campaigns with 'alter-globalization' movements engaged in the construction of alternative community economies and culture of water, centred on concepts such as the commons and 'water democracies'. In this third section of the paper, an attempt is made to complicate the public/private, commodity/rights, citizen/customer binaries underpinning much of the debate, through exploring the different socio-economic identities of citizens and different property rights, invoked under different water management models around the world. In the concluding section, the conceptual and political implications of this analysis are teased out, focusing on the implications of this analysis for our understandings of 'neoliberal natures'.

Neoliberal reforms and resource management: clarifying the debate

Much of the literature on 'neoliberalizing nature' is concerned with the creation of private property rights for resources previously governed as common pool resources. Of particular interest have been the impacts of 'neoliberalism' on specific resources (for edited collections, see Heynen et al, 2007; Mansfield, 2008; for water-specific studies see Bakker, 2004, 2010; Finger and Allouche, 2002; Johnstone and Wood, 2001; Maddock, 2004; McDonald and Ruiters, 2005; Perrault, 2006; Prasad, 2006; Shirley, 2002; Smith, 2004).

As Noel Castree notes in his review of this literature (Castree, 2005), much of this work has emphasized case-specific analyses of very different types of processes broadly grouped under the rather nebulous banner of neoliberalization: privatization, marketization, de-regulation, re-regulation, commercialization and corporatization, to name just a few. Although Castree acknowledges the utility of this work in illustrating that 'neoliberalism' is actually constituted of a range of diverse, locally rooted practices of neoliberalization, he identifies two analytical traps: failure to identify criteria by which different cases of neoliberalizing nature can be deemed sufficiently similar in order to conduct comparisons; and the occlusion of distinct types of neoliberal practices when subsumed under the broad (and overly general) label of neoliberalism. This paper responds to Castree's call for analytical frameworks with which to clarify these issues. As Sparke notes in a recent review (2006), this task is both analytically and politically crucial, insofar as the ideal types to which some of this work falls prey risk reinforcing or even reproducing the idealism of neoliberalism itself.

In developing such an analytical framework, an iterative approach is required which articulates (and revises) conceptual frameworks of neoliberalization (as a higher-order abstraction) and empirical analysis of the contingent mediation of neoliberal agendas by historically and geographically specific material conditions and power relations. In undertaking this analysis, it is important to distinguish between three categories of resource management upon which neoliberal reforms can be undertaken. Resource management *institutions* are the laws, policies, rules, norms and customs by which resources are governed. Resource management *organizations* are the collective social entities that govern resource use. And resource management *governance* is the process by which organizations enact management institutions; the practices by which, in other words, we construct and administer the exploitation of resources (Table 2.1).

As illustrated in Table 2.1, reforms can be undertaken in distinct categories and are not necessarily concomitant; one may privatize without de-regulating;

Table 2.1 Resource management reforms: examples from the water sector

Category	Target of reform	Type of reform	Example drawn from the water sector
Resource management institutions	Property rights	Privatization (enclosure of the commons or asset sale)	Introduction of riparian rights (England; Hassan, 1998); or sale of water supply of infrastructure to private sector (England and Wales; Bakker, 2004)
	Regulatory frameworks	De-regulation	Cessation of direct state oversight of water quality mechanisms (Ontario, Canada; Prudham, 2004)
Resource management organizations	Asset management	Private sector 'partnerships' (outsourcing contracts)	French municipal outsourcing of water supply system management to private companies (Lorrain, 1997)
	Organizational structure	Corporatization	Conversion of business model for municipal water supply: from local government department to a publicly owned corporation (Amsterdam, the Netherlands; Blokland et al, 2001)
Resource Governance	Resource allocation	Marketization	Introduction of a water market (Chile; Bauer, 1998)
	Performance incentives/sanctions	Commercialization	Introduction of commercial principles (e.g. full cost recovery) in water management (South Africa; McDonald and Ruiters, 2005)
	User participation	Devolution or decentralization	Devolving water quality monitoring to lower orders of government or individual water users (Babon River, Indonesia; Susilowati and Budiati, 2003)

de-regulate without marketizing; and commercialize without privatizing, etc. To give a simple example: privatization of the water supply industry in England and Wales in 1989 did not entail marketization; that is, it did not entail the introduction of markets in water abstraction licenses. This example illustrates one of the main confusions which arises in the literature: reforms to institutions, organizations and governance are all subsumed under the general term 'neoliberalization', despite the fact they often involve very different types of reforms, applied to different aspects of resource management. Another source of confusion arises when different types of reforms are assumed to be interchangeable, and when distinct terms (marketization, privatization) are assumed to be synonymous, when they are not.

How is such a typological exercise helpful in either analysis or activism? First, the failure to distinguish between categories of resource management and between targets and types of reforms obscures the specificity of the reform processes which are the object of analysis and limits our ability to compare cases, as Castree has noted. For example, comparing the introduction of water rights for 'raw' water (water in nature) (Haddad, 2000) in California to private sector participation in water supply management in New York (Gandy, 2004) is of limited interest, because two distinct processes are at work (marketization versus private sector participation). In contrast, comparing the introduction of water markets in Chile (Bauer, 1998) and California (Haddad, 2000) is worthwhile, because in both cases private property rights for water supply have been introduced via a process of marketization of water resource allocation. In short, the typology presented in Table 2.1 is analytically useful because it enables us to correctly compare different types of market environmentalist reforms and to more accurately characterize their goals and evaluate their outcomes.

This typology is also useful in addressing the widespread failure to adequately distinguish between different elements of neoliberal reform processes, an analytical sloppiness that diminishes our ability to correctly characterize the aims and trajectories of neoliberal projects of resource management reform (Bakker, 2005). Commercialization, for example, often precedes privatization in the water supply sector, which is sometimes followed by attempts to commodify water. The biophysical properties of resources, together with local governance frameworks, strongly influence the types of neoliberal reforms which are likely to be introduced: common-pool, mobile resources such as fisheries are more amenable to marketization, whereas natural monopolies such as water supply networks are more amenable to privatization (Bakker, 2004). In other words, in failing to exercise sufficient analytical precision in analysing processes of 'neoliberalizing nature', we are likely to misinterpret the reasons for, and incorrectly characterize the pathway of, specific neoliberal reforms.

As explored in subsequent sections of the paper, this typology may also be useful in clarifying activist strategies and in structuring our analyses of activism and advocacy. For example, in much of the literature on 'neoliberal nature' (and in many NGO and activist campaigning documents), water as a 'commodity' is contrasted to water as a 'human right'. Careful conceptualization of the neoliberalization of water demonstrates that this is misleading, insofar as the term 'commodity' refers to a property rights regime applicable to resources, and human rights is a legal category applicable to individuals. The more appropriate, but less widely used, antonym of water as a 'commodity' would more properly be a water 'commons'. As explored in the following sections, this distinction has had significant implications for the success of 'anti-privatization' and 'alter-globalization' struggles around the world.

Debating neoliberalization: anti-privatization campaigns and the 'human right to water'

The international campaign for a human right to water has grown enormously over the past decade. This campaign has its roots in the arguments of anti-privatization campaigners, who have fought numerous campaigns to resist and then overturn water privatization projects around the world. Advocates of private sector involvement in water supply – private companies, bilateral aid agencies and many governments – argue that it will increase efficiency and deliver water to those who currently lack access. They point to the failure of governments and aid agencies to achieve the goal of universal water supply during the International Water and Sanitation Decade (1981– 1990) and to the low efficiency and low levels of cost recovery of public utilities. Through efficiency gains and better management, private companies will be able to lower prices, improve performance and increase cost recovery, enabling systems to be upgraded and expanded, critical in a world in which one billion people lack access to safe, sufficient water supplies. Privatization (the transfer of ownership of water supply systems to private companies) and private sector 'partnerships' (the construction, operation and management of publicly owned water supply systems by private companies) have, it is argued, worked well in other utility sectors (see, for example, DFID, 1998; Dinar, 2000; Rogers et al, 2002; Shirley, 2002; Winpenny, 2003).

This view has been strongly critiqued by those who argue that neoliberalization entails an act of dispossession with negative distributive consequences that is emblematic of 'globalization from above' (Assies, 2003; Barlow and Clarke, 2003; Bond, 2004a; Hukka and Katko, 2003; McDonald and Ruiters, 2005; Petrella, 2001; Shiva, 2002). According to its opponents, the involvement of private companies invariably introduces a pernicious logic of the market into water management, which is incompatible with guaranteeing citizens' basic right to water. Private companies – answerable to shareholders and with the overriding goal of profit – will manage water supply less sustainably than public sector counterparts. Opponents of privatization point to successful examples of public water systems and research that private sector alternatives are not necessarily more efficient, and often much more expensive for users, than well-managed public sector systems (see, for example, Estache and Rossi, 2002). They assert the effectiveness of democratic accountability to citizens when compared with corporate accountability to shareholders; an argument less easy to refute following the collapse of Enron, which by the late 1990s had become one of the largest water multinationals through its subsidiary Azurix.

Opponents of water supply privatization frequently invoke a human right to water to support their claims (Gleick, 1998; Hukka and Katko, 2003; Morgan, 2004b, 2005; Trawick, 2003). The argument for creating a human right to water generally rests on two justifications: the non-substitutability of drinking water ('essential for life') and the fact that many other human rights which are explicitly recognized in the UN Conventions are predicated upon an (assumed) availability of water (e.g. the right to food).

The claim to a human right to water is not explicit in international law: no explicit right to water is expressed in the International Covenant on Economic, Social and Cultural Rights, and none of the United Nations conventions on human rights (except article 24 of the Convention on the Rights of the Child) explicitly recognizes the right to water (Morgan, 2004a). However, the UN Committee on Economic, Social and Cultural Rights issued a comment in 2002, asserting that every person has a right to 'sufficient, safe, acceptable, physically accessible, and affordable water' (ECOSOC, 2002; Hammer, 2004). Accordingly, a significant element of anti-privatization campaigning of NGOs in both the North and South has been a set of intertwined campaigns for the human right to water, beginning with a set of declarations by activists in both the North and the South (including the Cochabamba Declaration, the Group of Lisbon's Water Manifesto (Petrella, 2001) and the P8 Declaration in 2000), and growing to include well-resourced campaigns hosted by high-profile NGOs such as Amnesty International, the World Development Movement, the Council of Canadians, the Sierra Club, Jubilee South, Mikhail Gorbachev's Green Cross and Ralph Nader's Public Citizen. Activists have also focused on country-specific campaigns for constitutional and legal amendments, notably Uruguay's 2004 successful referendum resulting in a constitutional amendment creating a human right to water.

As the anti-water privatization campaign has transformed into a campaign for the human right to water, activists have gained support from mainstream international development agencies including the World Health Organization and the United Nations Development Programme (ECOSOC, 2002; UNDP, 2006; UN Economic and Social Council, 2003; WHO, 2003). These agencies articulate several arguments in favour of the human right to water: higher political priority given to water issues; new legal avenues for citizens to compel states to supply basic water needs; and the fact that the right to water is implicit in other rights (such as the rights to food, life, health and dignity) which have already been recognized in international law and which are implicitly recognized through legal precedents when courts support right of non-payment for water services on grounds of lack of affordability (UNWWAP, 2006).

Opponents have pointed out the difficulty of implementing a 'right to water': lack of clear responsibility and capacity for implementation; the possibility of causing conflict over transboundary waters; and potential abuse of the concept as governments could over-allocate water to privileged groups, at the expense of both people and the environment. Others argue that a right to water will effect little practical change: the right to access to water supply enshrined in South Africa's post-apartheid constitution, for example, has not prevented large-scale disconnections and persistent inequities in water distribution (Bond, 2002; McDonald and Ruiters, 2005). Another critique pertains to the anthropocentrism of human rights, which fail to recognize rights of non-humans (or ecological rights); providing a human right to water may, ironically, imply the further degradation of hydrological systems upon which we depend.

Another, more fundamental criticism is the argument that a human right to water does not foreclose private sector management of water supply systems. Critics of human rights doctrines argue that 'rights talk' stems from an individualistic, libertarian philosophy that is 'Eurocentric' (see, for example, Ignatieff, 2003; Kymlicka, 1995; Mutua, 2002; Rorty, 1993); as such, human rights are compatible with capitalist political economic systems. In other words, private sector provision is compatible with human rights in most countries around the world. A human right to water does not imply that water should be accessed free (although it might imply an affordable basic 'lifeline' supply) (UNWWAP, 2006), although this is at odds with cultural and religious views on water access in many parts of the world (for example, the definition of water as collective property ('waqf') under Islam (Faruqui et al, 2003)). Indeed, the UN's Committee on Economic, Social and Cultural Rights recognized the ambivalent status which a human right conveys upon a resource when it defined water as a social, economic and cultural good as well as a commodity (ECOSOC, 2003).

Many citizens of capitalist democracies accept that commodities are not inconsistent with human rights (such as food, shelter), but that some sort of public, collective 'safety net' must exist if these rights are to be met for *all* citizens. This is true for housing and food (as inadequate as these measures may be in practice). The situation with drinking water is more complicated, because drinking water is a non-substitutable resource essential for life and because networked water supply is a natural monopoly subject to significant environmental externalities. In this case, strong market failures provide an overwhelming justification for public regulation and, in many cases, owner-ship of assets. Full privatization is thus inconsistent with a human right to water unless it is coupled (as it is in England) with a universality requirement (laws prohibiting disconnections of residential consumers) and with strong regulatory framework for price controls and quality standards. Private sector participation in water supply, on the other hand, certainly fits within these constraints. In short rooted in a liberal tradition that prioritizes private ownership and individual rights, the current international human rights regime is flexible enough to be fully compatible with private property rights, whether for water or other basic needs.

In summary, pursuing a 'human right to water' as an anti-privatization campaign makes three strategic errors: conflating human rights and property rights; failing to distinguish between different types of property rights and service delivery models; and thereby failing to foreclose the possibility of increasing private sector involvement in water supply. Indeed, the shortcomings of 'human right to water' anti-privatization campaigns became apparent following the Kyoto World Water Forum, as proponents of private sector water supply management began speaking out in favour of water as a human right. Senior water industry representatives identified water as a human right on company websites, in the media, and at high-profile events such as the Davos World Economic Forum. Right-wing think tanks such as the Cato Institute backed up these statements with reports arguing that 'water socialism' had failed the poor and that market forces, properly regulated, were the best means of fulfilling the human right to water (Bailey, 2005; Segerfeldt, 2005). Non-governmental organizations such as the World Water Council, regarded by anti-privatization campaigners as being allied with private companies also developed arguments in favour of water as a human right (Dubreuil, 2005, 2006). Shortly after the Kyoto meeting, the World Bank released a publication acknowledging the human right to water (Salman and McInerney-Lankford, 2004).

Two years later, at the Fourth World Water Forum in Mexico City in 2006, representatives of private water companies issued a statement recognizing the right to water and recalling that the private sector had officially endorsed the right to water in 2005 at the 13th session of the UN's Commission on Sustainable Development (Aquafed, 2006). At the Mexico City Forum, a somewhat contrived consensus across civil society, the private sector and governments on the 'right to water' emerged (Smets, 2006). Despite dissenting views of Third World governments such as Bolivia, a 'diluted' interpretation of the human right to water prevailed in the Ministerial Declaration of the Fourth World Water Forum, in regards to which private companies had an officially sanctioned role.

Ironically, this has occurred at the same time as private companies have been acknowledging the significant barriers to market expansion in the water supply sector in the South. Analysis of the discourse of the public statements of senior executives of water supply services firms reveals a retreat from earlier commitments to pursuing PSPs globally, with senior figures publically acknowledging high risks and low profitability in supplying the poor (Robbins, 2003). Some international financial institutions have begun officially acknowledging the limitations of the private sector (ADB, 2003; UNDP, 2003). High-profile cancellations of water supply concession contracts – including Atlanta, Buenos Aires, Jakarta, La Paz and Manila – seem to bear out the hypothesis that water presents difficult, and perhaps intractable, problems for private sector management. The private sector has indeed retreated from supplying water to communities in the South, but this has been largely due to the failure to achieve acceptable return on investment and control risk, not to anti-privatization, pro-human rights campaigns. Companies continue to insist that water is a human right, which they are both competent and willing to supply, if risk-return ratios are acceptable, but this not a condition which cannot be met by most communities.

Alter-globalization and the commons

In reflecting on the failure of the 'human right to water' campaigns to foreclose the involvement of the private sector in water supply management, we broach a question often raised by 'alter-globalization' activists: how can we negotiate resistance to neoliberalization? In attempting to answer this question, activists are often divided on the question of human rights. Some argue that 'rights talk' resuscitates a public/private binary that recognizes only two unsatisfactory options – state or market control: twinned corporatist models from which communities are equally excluded (see, for example, Olivera and Lewis, 2004; Roy, 1999; Shiva, 2002). Instead, some activists argue in favour of alternative concepts of property rights, most frequently some form of the 'commons', to motivate their claims, juxtaposing this view to that of water as a commodity (Table 2.2).

At the risk of over-simplification, the commodity view asserts that private ownership and management of water supply systems (in distinction from water itself) is possible and indeed preferable. From this perspective, water is no different than other essential goods and utility services. Private companies, who will be responsive both to customers and to shareholders, can efficiently run and profitably manage water supply systems. Commercialization rescripts water as an economic good rather than a public good and redefines users as individual customers rather than a collective of citizens.

Table 2.2 The commons versus commodity debate

	Commons	Commodity
Definition	Public good	Economic good
Pricing	Free or 'lifeline'	Full-cost pricing
Regulation	Command and Control	Market based
Goals	Social equity and livelihoods	Efficiency and water security
Manager	Community	Market

Water conservation can thus be incentivized through pricing – users will cease wasteful behaviour as water prices rise with increasing scarcity. Proponents of the 'commodity' view assert that water should be treated like any other economic good – such as food – essential for life.

In contrast, the commons view of water asserts its unique qualities: water is a flow resource essential for life and ecosystem health; non-substitutable and tightly bound to communities and ecosystems through the hydrological cycle (Shiva, 2002; TNI, 2005). From this perspective, collective management by communities is not only preferable but also necessary, for three reasons. First, water supply is subject to multiple market *and* state failures; without community involvement, we will not manage water wisely. Second, water has important cultural and spiritual dimensions that are closely articulated with place-based practices; as such, its provision cannot be left up to private companies or the state. Third, water is a local flow resource whose use and health are most deeply impacted at a community level; protection of ecological and public health will only occur if communities are mobilized and enabled to govern their own resources. In particular, those who advance the 'commons' view assert that conservation is more effectively incentivized through an environmental, collectivist ethic of solidarity, which will encourage users to refrain from wasteful behaviour. The real 'water crisis' arises from socially produced scarcity, in which a short-term logic of economic growth, twinned with the rise of corporate power (and in particular water multinationals) has 'converted abundance into scarcity' (Shiva, 2002). As a response to the Dublin Principles, for example, the P7 Declaration (2000), outlined principles of 'water democracy', of decentralized, community-based, democratic water management in which water conservation is politically, socio-economically and culturally inspired rather than economically motivated.

Despite their divergent political commitments, opponents and proponents of neoliberalization of water supply share some common conceptual commitments, including an understanding (lacking in many 'neoliberalizing nature' analyses) that commodification is fraught with difficulty. In the language of regulatory economists and political scientists, water is conventionally considered to be an imperfect public good (non-excludable but rival in consumption) which is highly localized in nature and which is often managed as a common-pool resource, for which relatively robust community-controlled cooperation and management mechanisms exist in many parts of the world (Berkes, 1989; Mehta, 2003; Ostrom, 1990). It is the combination of public good characteristics, market failures and common property rights which makes water such an 'uncooperative' commodity and so resistant to neoliberal reforms, as neoclassical economists recognize when referring to the multiple 'market failures' that characterize resources such as water supply (Bakker, 2004). To rephrase this analysis in political ecological terms: water is a flow resource over which it is difficult to establish private property rights; is characterized by a high degree of public health and environmental externalities – the costs of which are difficult to calculate and reflect in water prices; and is a partially non-substitutable resource essential for life with important aesthetic, symbolic, spiritual and ecological functions which render some form of collective, public oversight inevitable. Private property rights can be established for water resources or water supply infrastructure, but full commodification does not necessarily and in fact rarely follows.

A high degree of state involvement, therefore, is usually found even in countries that have experimented heavily with neoliberal forms to water management. Here lies the second point of convergence between 'commodity' and 'commons' proponents: both neoliberal reformers and defenders of the 'commons' invoke dissatisfaction with centralized, bureaucratic state provision (cf Scott, 1998). Whereas over much of the 20th century, 'public good' would have been opposed to 'economic good' in defense of the state against private interests by anti-privatization activists, alter-globalization movements – such as ATTAC and the Transnational Institute – explicitly reject state-led water governance models (Shiva, 2002; TNI, 2005). In doing so, as explored below, they reinvigorate a tripartite categorization of service delivery which undermines the 'public/private' binary implicitly underlying much of the debate on neoliberalism more generally (Table 2.3).

As indicated in Table 2.3, significant differences exist between the public utility, commercial and community governance models, despite the fact that these models overlap to some degree in practice. One important distinction is the role of the consumer: a citizen, a customer, or a community member. Each role implies different rights, responsibilities and accountability mechanisms. Yet this tri-partite categorization tends to compartmentalize water supply into ideal types. In fact, many governments have chosen to create hybrid management models. Some have chosen, for example, to retain ownership while corporatizing water services, as in the Netherlands. In France, private-sector management of municipally owned water supply infrastructure via long-term management contracts is widespread. Other countries such as Denmark, with a long tradition of cooperative management of the local economy, prefer the coop model – provision by a non-profit users 'association in which local accountability is a key incentive'. Moreover, this tripartite classification is clearly inadequate when applied to the global South, where 'public' water supply systems often supply only wealthier neighbourhoods in urban areas, leaving poor and rural areas to self-organize through community cooperatives or informal, private, for-profit provision by water vendors, often at volumetric rates much higher than those available through the public water supply system. Indeed, most residents use multiple sources of water in the home and rely on a mix of networked and artisanal water supply sources, through both state and private sector delivery systems, using a combination of household piped network water connections, shallow and deep wells, public hydrants and water vendors for their water supply needs (see, for example, Swyngedouw, 2004). A public/private binary, even where

Table 2.3 Water supply delivery models: the cooperative, the state and the private corporation

		State	Market	Community
Resource management institutions	Primary goals	Guardian of public interest	Maximization of profit	Serve community interest
	Regulatory framework	Conformity with legislation/policy	Efficient performance	Effective performance
		Command and control	Market mechanisms	Community-defined goals (not necessarily consensus based)
	Property rights	Public (state) or private property	Private property	Public (commons) or private property
Resource management organizations	Primary decision-makers	Administrators, experts, public officials	Individual households, experts, companies	Leaders and members of community organizations
	Organizational structure	Municipal department, civil service	Private company, corporation	Cooperative, association/network
	Business models	Municipally owned utility	Private corporate utility	Community cooperative
Resource governance	Accountability mechanism	Hierarchy	Contract	Community norms
	Key incentives	Voter/ratepayer opinion	Price signals (share movements or bond ratings), customer opinion	Community opinion
	Key sanctions	Political process via elections, litigation	Financial loss, takeover, litigation	Livelihood needs, social pressure, litigation (in some cases)
	Consumer role	User and citizen	User and customer	User and community member
	Participation of consumers	Collective, top-down	Individualistic	Collective, bottom-up

it admits to the possibility of a third 'cooperative' alternative, is clearly insufficient for capturing the complexity of water provision in cities in the South (Swyngedouw, 2004). Alternative community economies of water do, in fact, already exist in many cities in the South (Table 2.4) and represent 'actually existing alternatives' to neoliberalism which activists have sought to interrogate, protect and replicate through networks such as the 'Blue Planet Project', 'Octubre Azul', World Social Fora (Ponniah, 2004; Ponniah and Fisher, 2003) and alternative 'world water fora'.

In opening up space for the conceptual acknowledgement of alternative community economies (cf Gibson-Graham, 2006), this tactic is to be welcomed. Yet caution is also merited, insofar as appeals to the commons run the risk of romanticizing community control. Much activism in favour of collective, community-based forms of water supply management tends to romanticize communities as coherent, relatively equitable social structures, despite the fact that inequitable power relations and resource allocation exist within communities (McCarthy, 2005; Mehta, 2001; Mehta et al, 2001). Although research has demonstrated how cooperative management institutions for water common-pool resources can function effectively to avoid depletion (Ostrom, 1990; Ostrom and Keohane, 1995), other research points to the limitations of some of these collective action approaches in water (Cleaver, 2000; Mehta, 2001; Mosse, 1997; Potanski and Adams, 1998; St Martin, 2005). Commons, in other words, can be exclusive and regressive, as well as inclusive and progressive (McCarthy, 2005). Indeed, the role of the state in encouraging redistributive models of resource management, progressive social relations and redistribution is more ambivalent than those making calls for a 'return to the commons' would perhaps admit.

Thus, the most progressive strategies are those that adopt a twofold tactic: reforming rather than abolishing state governance, while fostering and sharing alternative local models of resource management. In some instances, these alternative strategies tackle the anthropocentrism of neoliberalization (and 'human right to water' campaigns) directly, recognizing ecological as well as human needs, the latter being constrained through a variety of norms, whether scientifically determined 'limits', eco-spiritual reverence, or eco-puritan ecological governance. In other cases, they may make strange bedfellows with some aspects of neoliberal agendas, such as decentralization, through which greater community control can be enacted (Table 2.4).

These models are necessarily varied; no one model of water governance can be anticipated or imposed (cf Gibson-Graham, 2006). Rather, they build on local resource management and community norms, whether rural water users' customary water rights ('usos y costumbres') in the Andes (Trawick, 2003); revived conceptions of Roman 'res publica' and 'res commmuna' in Europe (Squatriti, 1998); or community norms of collective provision of irrigation in Indian 'village republics' (Shiva, 2002; Wade, 1988). In each instance, a place-specific model of what Indian activist Vandana Shiva terms

Table 2.4 Neoliberal reforms and alter-globalization alternatives

Category	Target of reform	Type of reform	Alter-globalization alternative
Resource management institutions	Property rights	Privatization	Mutualization (re-collectivization) of asset ownership (Wales; Bakker, 2004)
	Regulatory frameworks	De-regulation	Communal water rights in village 'commons' in India (Narain, 2006) Re-regulation by consumer-controlled NGOs such as 'Customer Councils' in England (Page and Bakker, 2005)
Resource management organizations	Asset management	Private sector 'partnerships'	Public–public partnerships (e.g. between Stockholm's water company (Stockholm Vatten) and water utilities in Latvia and Lithuania) (PSIRU, 2006) Water cooperatives in Finland (Katko, 2000)
	Organizational structure	Corporatization	Low-cost, community-owned infrastructure (e.g. Orangi Pilot Project, Pakistan; Zaidi, 2001)
Resource governance	Resource allocation	Marketization	Sharing of irrigation water based on customary law ('usos y costumbres') in Bolivia (Trawick, 2003)
	Performance incentives/ sanctions	Commercialization	Customer corporation (with incentives structured towards maximization of customer satisfaction rather than profit or share price maximization; Kay, 1996)
	User participation	Devolution or decentralization	Community watershed boards (Canada; Alberta Environment, 2003) Participatory budgeting (Porto Alegre, Brazil; TNI, 2005)

'water democracy' emerges, offering a range of responses to the neoliberalization agendas identified earlier in the paper. In other words, these 'really
existing' alter-globalization initiatives are a form of what Gibson-Graham
terms 'weak theory': deliberately organic, tentative, local, place-based and
(at least at the outset) modest.

'Weak', does not, however, imply 'insignificant'. These reforms are, of
course, necessarily local – because water is usually consumed, managed and
disposed of at a local scale. But they are nonetheless replicable and thus
represent potentially powerful 'actually existing alternatives' to neoliberalization. One example is the recent proliferation of 'public public partnerships',
in which public water supply utilities with expertise and resources (typically
in large cities in the North) are partnered with those in the South, or with
smaller urban centres in the North (PSIRU, 2005, 2006; Public Citizen, 2002;
TNI, 2005). Activists have actively promoted these strategies as a tactic of
resistance to water supply privatization initiatives, while acknowledging the
political pitfalls of promoting public–public partnerships in the wake of failed
private sector contracts, particularly the potential for such partnerships
to be promoted as a strategy for less profitable communities, allowing more
limited private sector contracts to 'cherry pick' profitable communities. Institutional support from multilateral agencies may soon be forthcoming, as
the newly commissioned UN Secretary General's Advisory Board on Water
and Sanitation has requested the UN support the creation of an international
association of public water operators. Encouraged by the UN Commission
on Sustainable Development's official acknowledgment of the importance
of promoting public–public partnerships (TNI, 2006; UNCSD, 2005) and by
specific campaigns by public water supply utilities – notably Porto Alegre –
governments in Argentina, Bolivia, Brazil, Indonesia, Holland, Honduras,
France, South Africa and Sweden have initiated public–public partnerships,
at times also entailing a radical restructuring of management–worker relationships within water supply utilities (TNI, 2006).

Conclusions

As explored in this paper, the adoption of human rights discourse by private
companies indicates its limitations as an anti-privatization strategy. Human
rights are an individualistic, anthropocentric, state-centric and compatible
with private sector provision of water supply; and as such, a limited strategy
for those seeking to refute water privatization. Moreover, 'rights talk' offers
us an unimaginative language for thinking about new community economies,
not least because pursuit of a campaign to establish water as a human right
risks reinforcing the public/private binary upon which this confrontation is
predicated, occluding possibilities for collective action beyond corporatist
models of service provision. In contrast, the 'alter-globalization' debate
opened up by disrupting the public/private binary has created space for the

construction of alternative community economies of water. These 'alter-globalization' proposals counterpose various forms of the commons to commodity-based property and social relations. Greater progressive possibilities would appear to be inherent in the call of alter-globalization activists for radical strategies of ecological democracy predicated upon calls to de-commodify public services and enact 'commons' models of resource management (see, for example, Bond, 2004a, 2004b; TNI, 2005).

How does a more refined understanding of neoliberalization, as outlined in the typology introduced at the outset of this paper, assist in this task? First, it enables activism to be more precise in its characterization of 'actually existing' neoliberalisms and thus to develop alternatives which have more political traction. For example, the 'commons' is an effective strategy for combating privatization because it correctly opposes a collective property right to private property rights. Second, in locating the application of neoliberalization in specific historically and geographically contingent contexts, it emphasizes what Sparke terms the '*dis*locatable' idealism of neoliberalism (Sparke, 2006), both through generating alternatives and through demonstrating how ostensibly neoliberal reforms may be congruent with other political agendas. In so doing, it enables us to see that neoliberalism is not monolithic – and that it creates political opportunities that may be progressive. For example, some neoliberal reforms may be congruent with the goals of alter-globalization activists – such as decentralization leading to greater community control of water resources. Third, it reminds us to pay attention to the multiplicity of reforms that typically occur when 'neoliberalizing nature', not all of which focus on property rights. Specifically, the typology presented in Table 2.1 allows us to refine our academic analyses and activist responses to different types of neoliberalization, which vary significantly, opening up the creation of a range of alternative community water economies (Table 2.4).

Many of these alternatives, it should be noted, are not produced in reaction to neoliberalization, but rather resuscitate or develop new approaches to governing the relationship between the hydrological cycle and socio-natural economies and polities. Some aspects of these reforms are congruent with a neoliberal agenda, but the work of alter-globalization activists reminds us that they need not be subsumed by neoliberalization. Rather, these reforms open up new political ecological and socio-natural relationships through which an ethic of care – for non-humans as well as humans – can be developed. As this paper has argued, this 'alter-globalization' agenda necessitates a refinement of our conceptual frameworks of neoliberalization, accounting for multiple modes of property rights and service provision. This conceptual reframing allows us both to accurately analyse neoliberalization in situ and also to generate politically progressive strategies with which to enact more equitable political ecologies – particularly if our definitions of prospective 'commoners' are porous enough to include non-humans.

KAREN BAKKER

Postscript

Once published, ideas often travel in unexpected ways. This is certainly the case with my 2007 article (reprinted in this chapter) on the question of water as a human right. The debate that this article has inspired – ranging across geography, development studies, anthropology, environmental studies and beyond – reflects the degree to which 'rights talk' is a potent yet ambivalent strategy.

Given that the article is critical of the human right to water, which has generated a certain degree of controversy, some of my main points bear repeating. First, my analysis was focused on the 'privatization debate' and explicitly developed its critique of the human right to water as an anti-privatization strategy. The central point of my critique – that privatization is legally compatible with the human right to water – was borne out by subsequent events, notably the reports of the UN Special Rapporteur on the human right to safe drinking water and sanitation, which stated that 'the human rights framework does not express a preference over models of service provision' and that 'human rights are neutral as to economic models, (OHCHR, 2010). The human right to water, in other words, does not legally preclude the involvement of private companies in the provision of water services, nor does it preclude full-cost pricing or a range of other management mechanisms associated with the involvement of private companies.

Second, I argued that the human right to water leaves many questions unanswered. Can a human right to water be practically realized? If so, what institutions and legal mechanisms might be required? Does 'rights talk' fail to incorporate (and perhaps undermine) the norms and commitments of community-controlled water supply systems? What are the environmental consequences of the human right to water and the broader implications of governing the environment with a concept that is so deeply anthropocentric? And, perhaps most importantly, is a human right to water sufficient to generate the social solidarity – both material and political – necessary for the equitable provision of water supply to both humans and non-humans? I argued, in 2007, that it could not.

Third, and related to the above point, my main argument in terms of political tactics still, I believe, holds: that the focus on *property rights* rather than human rights offers us more potential for politically progressive strategies. Effectively realizing human rights, in other words, requires the articulation of property rights, water rights and the human right to water. This implies, for example, explicitly linking water access and land tenure and integrating the governance of water use and land use practices.

The preceding points still, I believe, hold true. But I have modified my argument in some important ways – partly as a result of responses to

37

the article (some of which are explored in this book). I shared my analysis early on with key participants in the anti-privatization struggle; this inspired some mutual rethinking of the utility of the human right to water (both conceptually and politically).

Most importantly, I should emphasize my belief that defending and extending the human right to water is, in the current conjuncture, necessary (although not entirely satisfactory). The human right to water is a crucially useful tactic for those without access to legitimize their struggles not only for water, but also for human dignity. In this aspirational sense, the concept of human right to water is a valuable tool, because it compels the powerful to consider the redistributive politics embedded in their water use practices. In other words, human rights are not *the* solution, but are rather a strategy for creating the context in which claims for social and environmental justice can be pursued.

Moreover, a broad interpretation of human rights may be a useful conceptual (and political) antidote to what some scholars term 'elite capture' of the benefits of development. This arises because of a central conundrum underpinning the debate over the human right to water: while government agencies often recognize that unserved communities have legitimate claims to social services, they are frequently unable to provide those services and so deal with their claims on the basis of political expediency. In order to collectively apply informal pressure on governmental actors to meet their demands, these marginalized populations seek to define themselves as communities with a clear identity (such as 'slum dwellers' or 'water sellers'). Here, I agree with scholars like Appadurai, Chatterjee and Roy: the social or 'consumption' rights these groups demand are distinct from human rights, which focus on individual rights characteristic of Western liberal-rights frameworks. (Recent academic debates over social citizenship make a similar point.)

Finally, mobilizing around human rights can foster solidarity. But what kind of solidarity and with whom? Grounding claims of solidarity takes us into the messy terrain of property rights and water rights and into active struggles to reconfigure the respective roles of states, markets and communities. The most progressive politics, in my opinion, will seek new expressions of eco-social (or socio-natural) justice that move us away from anthropocentric, individualistic notions of human rights. A recent and suggestive example is the case of Bolivia, where politicians have been at the vanguard of the international movement in support of the human right to water, while simultaneously promoting environmental rights (notably the *Ley de Derechos de la Madre Tierra* and a revised *Ley del Agua*). The Bolivian case suggests that the irony of the growing international consensus on the human right to water is that it may lead eventually to give rise to the conditions for its transcendence.

Acknowledgement

This chapter (outside of the Postscript) was originally published as Bakker, K. (2007) 'The "commons" versus the "commodity": Alter-globalization, privatization, and the human right to water in the global South', *Antipode*, vol 39, no 3, pp430–455. Reprinted with permission.

References

ADB (2003) *Beyond Boundaries: Extending Services to the Urban Poor*, Asian Development Bank, Manila

Alberta Environment (2003) *Water for Life: Alberta's Strategy for Sustainability*, Alberta Environment, Edmonton, Canada

Anderson, T. and Leal, D. (2001) *Free Market Environmentalism*, Palgrave, New York, NY

Angel, D. (2000) 'The environmental regulation of privatized industry in Poland', *Environment and Planning C: Government and Policy*, vol 18, no 5, pp575–592

Aquafed (2006) Statement on the right to water and role of local governments by Gérard Payen, during the opening session of the World Water Forum. http://www.aquafed.org/pdf/WWF4-openingGPRTW-LocGovPc2006-03-16.pdf

Assies, W. (2003) 'David versus Goliath in Cochabamba: Water rights, neoliberalism, and the revival of social protest in Bolivia', *Latin American Perspectives*, vol 30, no 3, pp14–36

Bailey, R. (2005) 'Water is a human right: How privatization gets water to the poor', *Reason Magazine* 17 August, http://www.reason.com/news/show/34992.html

Bakker, K. (2004) *An Uncooperative Commodity: Privatizing Water in England and Wales*, Oxford University Press, Oxford

Bakker, K. (2005) 'Neoliberalizing nature? Market environmentalism in water supply in England and Wales', *Annals of the Association of American Geographers*, vol 95, no 3, pp542–565

Bakker, K. (2010) *Privatizing Water: Governance Failure and the World's Urban Water Crisis*, Cornell University Press, Ithaca

Barlow, M. and Clarke, T. (2003) *Blue Gold: The Fight to Stop the Corporate Theft of the World's Water*, Stoddart, New York, NY

Bauer, C. (1998) 'Slippery property rights: Multiple water uses and the neoliberal model in Chile, 1981–1995', *Natural Resources Journal*, vol 38, pp109–154

Berkes, F. (1989) *Common Property Resources: Ecology and Community-based Sustainable Development*, Belhaven Press, London

Bernstein, S. (2001) *The Compromise of Liberal Environmentalism*, Columbia University Press, New York, NY

Blokland, M., Braadbaart, O. and Schwartz, K. (2001) *'Private Business, Public Owners'*, Ministry of Housing, Spatial Planning, and Development and the Water Supply and Sanitation Collaborative Council, The Hague and Geneva

Bond, P. (2002) *Unsustainable South Africa: Environment, Development, and Social Protest*, Merlin Press, London

Bond, P. (2004a) 'Water commodification and decommodification narratives: Pricing and policy debates from Johannesburg to Kyoto to Cancun and back', *Capitalism Nature Socialism*, vol 15, no 1, pp7–25

Bond, P. (2004b) 'Decommodification and deglobalisation: Strategic challenges for African social movements', *Afriche e Oriente*, vol 7, no 4

Castree, N. (2005) 'The epistemology of particulars: Human geography, case studies, and "context"', *Geoforum*, vol 36, no 5, pp541–666

Cleaver, F. (2000) 'Moral ecological rationality, institutions, and the management of common property resources', *Development and Change*, vol 31, no 2, pp361–383

DFID (1998) *Better Water Services in Developing Countries: Public–Private Partnership – The Way Ahead*, Department for International Development, London

Dinar, A. (2000) *The Political Economy of Water Pricing Reforms*, World Bank, Washington, DC

Dubreuil, C. (2005) *The Right to Water: From Concept to Implementation*, World Water Council

Dubreuil, C. (2006) *Synthesis on the Right to Water at the 4th World Water Forum, Mexico*, World Water Council

ECOSOC (2002) 'General comment 15', United Nations Committee on Economic, Social and Cultural Rights, Geneva

Estache, A. and Rossi, C. (2002) 'How different is the efficiency of public and private water companies in Asia?', *World Bank Economic Review*, vol 16, no 1, pp139–148

Faruqui, N., Biswas, A. and Bino, M. (2003) *La gestion de l'eau selon l'Islam*, CRDI/Editions, Karthala

Finger, M. and Allouche, J. (2002) *Water Privatisation: Transnational Corporations and the Re-regulation of the Water Industry*, Spon Press, London

Frérot, A. (2006) 'Parce que ce droit est fondamental, il doit devenir effectif', *Le Monde*, 17 March

Gandy, M. (2004) 'Rethinking urban metabolism: Water, space and the modern city', *City*, vol 8, no 3, pp363–379

Gibson-Graham, J. K. (2006) *A Postcapitalist Politics*, University of Minnesota Press, Minneapolis, MN

Glassman, J. (2006) 'Primitive accumulation, accumulation by dispossession, accumulation by "extra-economic" means', *Progress in Human Geography*, vol 30, no 5, pp608–625

Gleick, P. (1998) 'The human right to water', *Water Policy*, vol 1, pp487–503

Goldman, M. (2005) *Imperial Nature: The World Bank and the Making of Green Neoliberalism*, Yale University Press, New Haven, CT

Haddad, B. (2000) *Rivers of Gold: Designing Markets to Allocate Water in California*, Island Press, Washington, DC

Hammer, L. (2004) 'Indigenous peoples as a catalyst for applying the human right to water', *International Journal on Minority and Group Rights*, vol 10, pp131–161

Hassan, J. (1998) *A History of Water in Modern England and Wales*, Manchester University Press, Manchester, UK

Heynen, N., McCarthy, J., Prudham, W. S. and Robbins, P. (eds) (2007) *Neoliberal Environments: False Promises and Unnatural Consequences*, Routledge, New York, NY

Hukka, J. J. and Katko, T. S. (2003) 'Refuting the paradigm of water services privatization', *Natural Resources Forum*, vol 27, no 2, pp142–155

Ignatieff, M. (2003) *Human Rights as Politics and Idolatry*, Princeton University Press, Princeton, NJ

Johnston, J., Gismondi, M. and Goodman, J. (2006) *Nature's Revenge: Reclaiming Sustainability in an Age of Corporate Globalization*, Broadview Press, Toronto

Johnstone, N. and Wood, L. (2001) *Private Firms and Public Water. Realising Social and Environmental Objectives in Developing Countries*, International Institute for Environment and Development, London

Katko, T. (2000) *Water! Evolution of Water Supply and Sanitation in Finland from the mid-1800s to 2000*, Finnish Water and Waste Water Works Association, Tampere, Finland

Kay, J. (1996) 'Regulating private utilities: The customer corporation', *Journal of Cooperative Studies*, vol 29, no 2, pp28–46

Kymlicka, W. (1995) *Multicultural Citizenship: A Liberal Theory of Minority Rights*, Oxford University Press, Oxford, UK

Laxer, G. and Soron, D. (2006) *Not for Sale: Decommodifying Public Life*, Broadview Press, Toronto

Lorrain, D. (1997) 'Introduction—the socio-economics of water services: The invisible factors' in Lorrain, D. (ed.) *Urban Water Management—French Experience Around the World*, Hydrocom, Levallois-Perret, France

Maddock, T. (2004) 'Fragmenting regimes: How water quality regulation is changing political-economic landscapes', *Geoforum*, vol 35, no 2, pp217–230

Mansfield, B. (ed.) (2008) *Privatization: Property and the Remaking of Nature-Society Relations*, Blackwell, Oxford

McCarthy, J. (2005) 'Commons as counter-hegemonic projects', *Capitalism Nature Socialism*, vol 16, no 1, pp9–24

McCarthy, J. and Prudham, S. (2004) 'Neoliberal nature and the nature of neoliberalism', *Geoforum*, vol 35, no 3, pp 275–283

McDonald, D. and Ruiters, G. (2005) *The Age of Commodity: Water Privatization in Southern Africa*, Earthscan, London

Mehta, L. (2001) 'Water, difference, and power: Unpacking notions of water "users" in Kutch, India', *International Journal of Water*, vol 1, pp3–4

Mehta, L. (2003) 'Problems of publicness and access rights: Perspectives from the water domain' in Kaul, I., Conceicao, P., Le Goulven, K. and Mendoza, R. (eds) *Providing Global Public Goods: Managing Globalization*, Oxford University Press and United Nations Development Program, New York, NY

Mehta, L., Leach, M. and Scoones, I. (2001) 'Editorial: Environmental governance in an uncertain world', *IDS Bulletin*, vol 32, no 4

Morgan, B. (2004a) 'The regulatory face of the human right to water', *Journal of Water Law*, vol 15, no 5, pp179–186

Morgan, B. (2004b) 'Water: frontier markets and cosmopolitan activism', *Soundings: a Journal of Politics and Culture*, vol 27, pp10–24

Morgan, B. (2005) 'Social protest against privatization of water: Forging cosmopolitan citizenship?' in Cordonier Seggier, M. C. and Weeramantry, J. (eds), *Sustainable Justice: Reconciling International Economic, Environmental and Social Law*, Martinus Nijhoff, The Hague

Mosse, D. (1997) 'The symbolic making of a common property resource: History, ecology and locality in a tank-irrigated landscape in South India', *Development and Change*, vol 28, no 3, pp467–504

Mutua, M. (2002) *Human Rights: A Political and Cultural Critique*, University of Pennsylvania Press, Philadelphia, PA

Narain, S. (2006) 'Community-led alternatives to water management: India case study', *Background Paper: Human Development Report 2006*, United Nations Development Programme, New York, NY

OHCHR (2010) Report of the independent expert on the issue of human rights obligations related to access to safe drinking water and sanitation. A/HRC/15/31, Office of the United Nations High Commissioner for Human Rights, Geneva

Olivera, O. and Lewis, T. (2004) *Cochabamba! Water War in Bolivia*, South End Press, Boston, MA

Ostrom, E. (1990) *Governing the Commons: The Evolution of Institutions for Collective Action*, Cambridge University Press, New York, NY

Ostrom, E. and Keohane, R. (eds) (1995) *Local Commons and Global Interdependence: Heterogeneity and Cooperation in Two Domains*, Harvard University, Centre for International Affairs, Cambridge, MA

Page, B. and Bakker, K. (2005) 'Water governance and water users in a privatized water industry: Participation in policy-making and in water services provision – a case study of England and Wales', *International Journal of Water*, vol 3, no 1, pp38–60

Perrault, T. (2006) 'From the Guerra del Agua to the Guerra del Gas: Resource governance, popular protest and social justice in Bolivia', *Antipode*, vol 38, no 1, pp150–172

Petrella, R. (2001) *The Water Manifesto: Arguments for a World Water Contract*, Zed Books, London and New York, NY

Ponniah, T. (2004) 'Democracy vs. empire: Alternatives to globalization presented at the World Social Forum', *Antipode*, vol 36, no 1, pp130–133

Ponniah, T. and Fisher, W. F. (eds) (2003) *Another World is Possible: Popular Alternatives to Globalization at the World Social Forum*, Zed Press, New York, NY

Potanski, T. and Adams, W. (1998) 'Water scarcity, property regimes, and irrigation management in Sonjo, Tanzania', *Journal of Development Studies*, vol 34, no 4, pp86–116

Prasad, N. (2006) 'Privatisation results: Private sector participation in water services after 15 years', *Development Policy Review*, vol 24, no 6, pp669–692

Prudham, W. S. (2004) 'Poisoning the well: Neoliberalism and the contamination of municipal water in Walkerton, Ontario', *Geoforum*, vol 35, no 3, pp343–359

PSIRU (2005) *Public–Public Partnerships in Health and Essential Services*, University of Greenwich, Public Services International Research Unit

PSIRU (2006) *Public–Public Partnerships as a Catalyst for Capacity Building and Institutional Development: Lessons from Stockholm Vatten's Experience in the Baltic Region*, University of Greenwich, Public Services International Research Unit

Public Citizen (2002) *Public-Public Partnerships: A Backgrounder on Successful Water/Wastewater Reengineering Programs*, Public Citizen and Food and Water Washington Watch, Washington, DC

Robbins, P. (2003) 'Transnational corporations and the discourse of water privatization', *Journal of International Development*, vol 15, pp1073–1082

Rogers, P., de Silva, R., et al (2002) 'Water is an economic good: How to use prices to promote equity, efficiency, and sustainability', *Water Policy*, vol 4, no 1, pp1–17

Rorty, R. (1993) 'Human rights, rationality, and sentimentality' in Shute, S. and Hurley, S. (eds) *On Human Rights: The Oxford Amnesty Lectures*, Basic Books, New York, NY

Roy, A. (1999) *The Cost of Living*, Modern Library, London

Salman, S. and McInerney-Lankford, S. (2004) *The Human Right to Water: Legal and Policy Dimensions*, World Bank, Washington DC

Scott, J. (1998) *Seeing like a State: How Certain Schemes to Improve the Human Condition have Failed*, Yale University Press, New Haven, CT

Segerfeldt, F. (2005) *Water for Sale: How Business and the Market Can Resolve the World's Water Crisis*, Cato Institute, London

Shirley, M. (2002) *Thirsting for Efficiency*, Elsevier, London

Shiva, V. (2002) *Water Wars: Privatization, Pollution and Profit*, Pluto Press, London

Smets, H. (2006) 'Diluted view of water as a right: 4th World Water Forum', *Environmental Policy and Law*, vol 36, no 2, pp88–93

Smith, L. (2004) 'The murky waters of the second wave of neoliberalism: Corporatization as a service delivery model in Cape Town', *Geoforum*, vol 35, no 3, pp375–393

Sparke, M. (2006) 'Political geography: Political geographies of globalization (2) – governance', *Progress in Human Geography*, vol 30, no 3, pp357–372

Squatriti, P. (1998) *Water and Society in Early Medieval Italy, A.D. 400–1000*, Cambridge University Press, Cambridge, UK

St Martin, K. (2005) 'Disrupting enclosure in the New England fisheries', *Capitalism Nature Socialism*, vol 16, no 1, pp63–80

Susilowati, I. and Budiati, L. (2003) 'An introduction of co-management approach into Babon River management in Semarang, Central Java, Indonesia', *Water Science & Technology*, vol 48, no 7, pp173–180

Swyngedouw, E. (2004) *Social Power and the Urbanization of Water*, Oxford University Press, Oxford, UK

TNI (2005) *Reclaiming Public Water: Achievements, Struggles and Visions from Around the World*, Transnational Institute, Amsterdam

TNI (2006) *Public Water for All: The Role of Public–Public Partnerships*, Transnational Institute and Corporate Europe Observatory, Amsterdam

Trawick, P. (2003) 'Against the privatization of water: An indigenous model for improving existing laws and successfully governing the commons', *World Development*, vol 31, no 6, pp977–996

UNCSD (2005) *Report on the Thirteenth Session*, UN Commission on Sustainable Development, New York, NY E/CN.17/2005/12

UNDP (2003) *Millennium Development Goals: A Compact for Nations to End Human Poverty*, United Nations Development Program, New York, NY

UNDP (2006) *Beyond Scarcity: Power, Poverty, and the Global Water Crisis: UN Human Development Report 2006*, United Nations Development Programme, Human Development Report Office, New York, NY

UN Economic and Social Council (2003) *Economic, Social and Cultural Rights. Report submitted to the 59th session of the Commission on Human Rights, by the Special Rapporteur on the Right to Food*, E/CN.4/2003/54

UNWWAP (2006) *Water: A Shared Responsibility*, United Nations World Water Assessment Program, New York, NY

Wade, R. (1998) *Village Republics: Economic Conditions for Collective Action in South India*, Cambridge University Press, Cambridge, UK

WHO (2003) *The Right to Water*, World Health Organisation, Geneva

Winpenny, J. (1994) *Managing Water as an Economic Resource*, Routledge, London

Winpenny, J. (2003) *Financing Water for All: Report of the World Panel on Financing Water Infrastructure*, World Water Council/Global Water Partnership/Third World Water Forum, Geneva

Zaidi, A. (2001) *From Lane to City: The Impact of the Orangi Pilot Project's Low Cost Sanitation Model*, WaterAid, London

3

THE HUMAN RIGHT
TO WHAT?

Water, rights, humans, and the relation of things

Jamie Linton

In this chapter the idea of the human right to water is approached from a relational perspective. The fluidity of water and humanity are taken as a starting point for considering how, when mediated by the idea of a "right", a kind of relation is expressed that entails the co-production of both. When considered in this way, proclaiming a specific human right to water fixes humans and water in a certain kind of relationship. Moreover, by redefining the right, the identity of both "human" and "water" are changed. The chapter proposes broadening the idea of the human right to water from this relational perspective. As it is usually stated, this right defines a relationship between the individual human being and a certain quantity of water necessary for personal and domestic needs. One may however retain the rhetorical, political and potentially legal power of the human right to water while formulating the right in a way that allows for a diversity of hydrosocial arrangements. Building on the idea that a right constitutes a kind of relation, it is theoretically and practically possible to call for rights that involve, on one hand, a collective identity of human being – or to use Marx's term, "species being" – and on the other hand an identity of water as a process rather than a quantity. Reformulating the human right to water thus rests on and gives strength to identities of humanity and water that are collective, processual and interrelated.

Two specific ideas are developed below: First, that the human right to water may be formulated so as to define the right to be involved in decisions that affect the way water and people articulate in the hydrosocial process. This formulation of the human right to water establishes a rule of governance. Second, the chapter considers how the human right to water might comprise the right of the collective to a portion of the productive capacity of water, or a portion of the economic value that is generated by means of setting water to the production of commodities. The principle advocated here is that the contribution of water to the value of economic production constitutes a right that should redound to the benefit of society generally. This right establishes a rule of social equity. Neither of these ideas is new – they have been recognized and practiced in different ways – however by formulating them as a human right they may be made compelling and given a prominence that would help advance a progressive political agenda.

The chapter begins, in the following section, with a discussion of the idea of the human right to water as it has been developed in international discourse, showing how this right is defined as a quantity of water necessary to meet the personal and domestic needs of the individual. Following a discussion of recent objections to propounding this right as an effective strategy for promoting social justice and welfare, a relational definition of the right to water is proposed. Such a definition allows us to retain a demand for the human right to water while formulating this demand in terms of rules of governance and of social equity.

The human right to water and its critics

The idea of the human right to water has been expressed in various international forums and documents as well as by academics and activists over the past three decades. For the most part, this right has been formulated in terms of specific quantities of water for personal needs. The declaration of the United Nations Water Conference held at Mar Del Plata, Argentina in 1977, often considered the first proclamation of the right to water, held that "all peoples . . . have the right to have access to drinking water in quantities and of a quality equal to their basic needs." (quoted in United Nations Department of Economic and Social Affairs, 1992) Since then, virtually every international statement on water supply (and sanitation) has reflected the notion of the human right to a quantity of water. Arguing for formal adoption of this right in international law in the late 1990s, Peter Gleick advocated a measure of 50 litres of clean water per person per day, a standard he deemed adequate to the basic needs of an individual, independent of climate, technology and culture (Gleick, 1996). While various authorities have put forward slightly different standards, most would agree with Gleick that "the specific number is less important than the principle of setting a goal and implementing actions to reach that goal." (Gleick, 1998, p496)

The instrument in international law providing the strongest juridical basis for the human right to water is the International Covenant on Economic, Social and Cultural Rights, adopted by the United Nations General Assembly in 1966 (Irujo, 2007, p268). As mentioned in the chapter by Sultana and Loftus in this book, in 2002 the Committee on Economic, Social and Cultural Rights (subsidiary to the International Covenant) published "General Comment

Number 15 – The Right to Water" which, for the first time explicitly recognized the human right to water. This General Comment specifically noted "An adequate amount of safe water is necessary to . . . provide for consumption, cooking, personal and domestic hygienic requirements." (United Nations Economic and Social Council, 2003, pp1–2) The July 2010 United Nations General Assembly resolution recognizing the human right to water was shortly followed by a resolution of the United Nations Human Rights Council affirming the right as legally binding and once again defining this right in terms of ensuring "a regular supply of safe, acceptable, accessible and affordable drinking water and sanitation services of good quality and sufficient quantity." (United Nations General Assembly – Human Rights Council, 2010)

While commitment to water as a human right remains strong among human rights advocates and many NGOs, the concept has not been without criticism in the last few years. Karen Bakker, for example, has drawn attention to the compatibility of human rights and private sector involvement in water supply (Bakker, 2007 and chapter in this book; see also Irujo, 2007, pp273–274). Bakker points out that proponents of water privatization and private water companies have taken up the cause of water as a human right (Bakker, 2007, pp439–440, and chapter in this book). The root of the problem, Bakker argues, is that human rights doctrine is rooted in political and economic individualism. As a means of resisting neoliberalization of the water sector, she has instead called for a shift in tactics towards recognizing and empowering alternative concepts of property rights, most notably in the form of common property water regimes, or "community economies of water".

Others have pointed out the vast range of actual and possible political and economic arrangements by which communities might secure for themselves rights of access to water as well as the power to be involved in decisions concerning water, its allocation and disposition. As Rajendra Pradhan and Ruth Meinzen-Dick have written, "rather than seeking a single, hegemonic type of water law or valuation of water, recognizing the pluralistic legal frameworks, types of rights, and meanings of water is not only a more realistic viewpoint [than arguing for the right to water as means of opposing privatization], but also one which can lead to more productive negotiations over water rights and water use." (Pradhan and Meinzen-Dick, 2010, p40)

Still others have argued that attention needs to shift away from a fixation on the human right to water toward the question of how such a right is to be realized through processes of governance. "Notably", argues David Brooks, "rights must be defined not only as so much water at such quality but also in the ability to participate in decisions about the delivery of that right." (Brooks, 2007, p238) As put by Amit Srivastava, executive director of the New Delhi-based India Resource Center, "For us, the right to water means the community has control over its water resources." (quoted in Bowe, 2009) By means of such arguments, the object of the right to water shifts from definite quantities of water to the capacity of people to be involved in water governance (see also Biswas, 2007).

Brooks and others have argued that the concept of the human right to water should also be broadened to include making water available for small farmers and householders to produce food and ensure ecosystem health (Brooks, 2007, p238). Asit Biswas contends, "If the concept of water as a human right is to progress further, it will be essential to consider other water uses . . . The debate and discussions need to cover a much wider territory." (Biswas, 2007, p220) In addition to the production of food, the assurance of environmental hygiene and the enjoyment of certain cultural practices, Biswas suggests this "territory" ought to include "other rapidly evolving water uses like water for energy production." (ibid.)

In the following two sections, I argue that these calls for recognizing community economies of water, encouraging participation in water governance and expanding the right to water to include broader aspects of production may be developed through a relational approach to the question of the right to water. This would have the effect of expanding and reformulating the idea of the human right to water.

Rights and relations: the right to water as a rule of governance

The rhetorical, political and potentially legal power of the human right to water might be retained while formulating this right in a different way: A right may be considered as a kind of relation, one that mediates between and defines the subject and object of the right in question. The human right to water, as usually stated, fixes a relation between the individual human body and a quantity of water. This formulation also establishes a certain relation between people in respect of water, namely as a collection of individuals. Stated in this way, such a right ignores the social nature of humanity as well as the processual nature of water. Building on the idea that a right constitutes a relation, we might consider the right to water as involving, on one hand, a collective identity of human being as species-being (Marx, 1978 [1843], p43) and on the other hand the identity of water as a process rather than a quantity (Linton, 2010, pp4–5). Defining the right to water in these terms is an argument for collective decision making in respect of the water process, thus producing a rule of governance. Such a definition responds to objections that have been raised by critics, as discussed in the previous section, suggesting the need to go beyond the limitations imposed by the fixed notion of the right to water and providing a more imaginative language for thinking about economies and cultures of water.

A relational approach holds that things become what they are in relation to other things that emerge through a process of mutual becoming (Linton, 2010, pp24–44). Accordingly, the human right to water defines a relation in which the identity of both "human" and "water" are established and sustained. From a political-ecology perspective, the usual formulation of the human right to water is conservative in the sense that it accomplishes little beyond sustaining the individual as a healthy organism. This sort of conservativism is characteristic of all human rights predicated on the salience of the individual human being. Here, Marx's critique of the classic "Rights of Man" is relevant to the argument. As Marx argued in 1843:

> None of the supposed rights of man, therefore, go beyond the egoistic man, man as he is, as a member of civil society; that is, an individual separated from the community, withdrawn into himself, wholly preoccupied with his private interest and acting in accordance with his private caprice. Man is far from being considered, in the rights of man, as a species-being; on the contrary, species-life itself – society – appears as a system which is external to the individual and as a limitation of his original independence.
>
> (Marx, 1978 [1843], p43)

As usually formulated, the human right to water is also conservative in respect of the identity of water as a substance of fixed quantity. A fixed quantity of water has a very restricted social life. Water, however, may be understood as a process as well as a substance. As every student of the hydrologic cycle knows, water is constitutionally on the move. As a force of nature as well as a factor of human history, it is the processual reality of water that defines its significance. The individual right to water may be understood in this sense, i.e. as one's right to claim a portion of this process for one's own drinking, cooking, cleaning, bathing and sanitation purposes. However, once considered in this way, the sovereignty of the individual is necessarily compromised by water's sociality, or what might be called the social effect of the water process (Hamlin, 2010). To drink, cook, clean, bathe or dispose of human waste with the aid of water cannot be regarded as a purely private matter. Water's constitutional mobility determines our every involvement with water as a social fact, as the depletions and adulterations we make to the water process in one way or another impact other people as well as hydrological processes themselves. Considered in this way, the right to water entails a relation between our species-being on one hand and the processual nature of water on the other.

Such thinking is conducive to arguments for broader participation in decision making about the water process. Considered in this way, the human right to water defines a rule of governance. This interpretation of the right to water is already present in human rights law. For example, Article 11 of General Comment 15 of 2002 introduced above states:

> The elements of the right to water must be adequate for human dignity, life and health, in accordance with articles 11, paragraph 1,

and 12. [of the International Convention on Economic, Social and Cultural Rights] The adequacy of water should not be interpreted narrowly, by mere reference to volumetric quantities and technologies. Water should be treated as a social and cultural good, and not primarily as an economic good. The manner of the realization of the right to water must also be sustainable, ensuring that the right can be realized for present and future generations.

(United Nations Economic and Social Council, 2003, p5)

Treating water "as a social and cultural good, and not primarily as an economic good" implies that the basic questions of how water is allocated and managed should be decided by democratic processes rather than market principles. Thus "the manner of the realization of the right to water" becomes an aspect of the very right itself. Here, the question of the right to water articulates with the concept of water governance. "Water governance", as defined by Karen Bakker and Linda Nowlan, "is the range of political, organizational, and administrative processes through which communities articulate their interests, their input is absorbed, decisions are made and implemented, and decision makers are held accountable in the development and management of water resources and delivery of water services." (Nowlan and Bakker, 2010, p7) Leaving aside for the moment the question what is meant by "communities", the key characteristic of the discursive and structural shift from "government" to "governance" involves an expanded role for nongovernmental actors in decision making with respect to water management and water services (Bakker, 2010, 44; deLoë and Kreutzwiser, 2007, p87). Simply put, the right to water thus recognizes the collective rights to be involved in determining the water process.

Elsewhere, Bakker has considered the failure of both public and private urban water systems to provide adequate services for the urban poor in terms of "governance failure", showing "how the institutional dimensions of water management and decision making do not effectively take into account the needs of all citizens." (Bakker, 2010, p45) While it mustn't be regarded as a panacea for the world's water problems, respecting peoples' rights to participate in decision making about water may be considered as complementary – and possibly prerequisite – to realizing the conventional form of the human right to water.

There remains, however, the difficulty of defining the "community" for purposes of interest articulation, and identifying the scale at which this right to participate in water decisions might best be realized. Water governance is usually understood as involving a rescaling of decision-making processes from the state towards smaller spatial and administrative units that are thought to be more conducive to accommodating peoples' needs and interests. Often the watershed, or river basin, has been presumed to be the most appropriate scale at which to organize processes of water governance (Molle,

2009; Nowlan and Bakker, 2010, pp8–9). These presumptions however, are fraught with practical and theoretical difficulties. As Emma Norman and Karen Bakker point out, while there has been "a significant increase in local water governance activities" in some places, "this has not resulted in a significant increase in decision-making power at the local scale." (Norman and Bakker, 2009, p99) Rescaling water governance to smaller units, moreover, does not necessarily produce more equitable decision-making processes or improved accountability (ibid., p104). As Bakker has shown, "community water management" is "not necessarily equitable or democratic" and can indeed be "exclusive and regressive, as well as inclusive and progressive." (Bakker, 2010, pp183–184) For this reason, she has argued "The need to balance equity and sustainability suggests the need for the continued, active role of the state in setting and enforcing water management criteria in community-managed initiatives." (ibid., p179)

Nor is the watershed/river basin necessarily the "natural" unit for water governance that it is often presumed to be. As Francois Molle has made clear, beyond its hydrological definition, the river basin is also "a political and ideological construct" that is no less a product of political and social dynamics than any other scalar configuration (Molle, 2009) (see also Bloomquist and Schlager, 2005). In a theoretical sense there is nothing about the community or the watershed that makes it the obvious or inherently preferable scale of water governance. Presuming the environmental, democratic and social justice benefits of rescaling processes of environmental governance to the local level has been characterized by J. Christopher Brown and Mark Purcell in terms of the "local trap" (Brown and Purcell, 2005). Drawing from the critique of scale that emerged in political economy in the 1990s, they point out that the "local", as with any other scale, is not given but produced in wider processes of social relations (Smith, 1992; Swyngedouw, 1997).

For purposes of the present argument, the right to water as a rule of governance should be understood to apply generally, and at all levels of decision making. When it comes to water governance, there are numerous decision-making processes operating simultaneously at a variety of co-related scales. In the case of water resources in Canada for example, the salient political processes operate at levels including the global, bi-national (with respect to boundary waters shared with the United States), nation-state, provincial state, and municipality, as well as the watershed scale. In accordance with the produced nature of scale, realizing the right to water as a rule of governance is bound to result in the production or privileging of certain scalar configurations. It thus has the potential of helping empower local actors' involvement in the water process by strengthening the case for the "commons" and for "community economies of water" that have been identified as "the most effective alternative to neoliberalism . . ." (Bakker, 2007, p444, and chapter in this book) Similarly, it would help facilitate what Erik Swyngedouw describes as an "emancipatory water politics" to challenge uneven power relations by means of "closer involvement of the local people in the management of and control over the public water utility." (Swynge-douw, 2004, p177)

By considering the human right to water as a rule of governance, a relational and dialectical process is suggested by which the right to water entails a change in social relations. Whereas the right to a certain quantity of water for the purposes of personal health and hygiene fixes social relations as a constellation of individuals (whose rights may be met by gaining access to the commodity form of water), the right to involvement in the water process dissolves individuality and refigures the rights-bearing entity as a collective whose interests can only be met through collective action. Such action has the corollary of changing society through the dialectical effect of the hydro-social process. The notion of changing ourselves through our engagements with water constitutes a kind of right akin to David Harvey's interpretation of the concept of the "right to the city" (see chapters by Sultana and Loftus, and by Bond in this book):

> The right to the city is far more than the individual liberty to access urban resources: it is a right to change ourselves by changing the city. It is, moreover, a common rather than an individual right since this transformation inevitably depends upon the exercise of a collective power to reshape the processes of urbanization. The freedom to make and remake our cities and ourselves is, I want to argue, one of the most precious yet most neglected of our human rights.
>
> (Harvey, 2008, p23)

The provision of water services in remote Indigenous Peoples' (First Nations, Inuit and Métis) communities in Canada provides an example of the need and the potential for considering the right to water in this relational and dialectical sense (Linton, 2010, pp225–227). Despite long-standing recognition of the right of people living in these communities to supplies of safe drinking water and despite considerable recent increases in government expenditures to this effect, the actual condition of drinking water remains deplorable: As of the spring of 2010, 114 First Nations communities in Canada were under Drinking Water Advisories – i.e. the water was unfit to drink – and 49 First Nations water systems were classified as "high risk" (Assembly of First Nations, 2010).

While inadequate financial resources and the lack of a proper legal regime have been identified as largely responsible for the poor state of drinking water services in First Nations communities (Phare, 2009, pp10–13), the problem can also be seen as a failure of local involvement in determining the hydrosocial process: The water supply and treatment technologies in these communities have been imported from places that have very different political cultures, knowledges, attitudes, tastes and types of expertise with respect to water. These water technologies typically have been developed in large urban centers that have little in common with remote First Nations communities. The lack of community acceptance of these drinking water systems and lack of local capacity to maintain them is a key reason for their failure (Department of Indian Affairs and Northern Development, 2006, 32; c.f. Safe Drinking Water Foundation, 2006).

In order to function effectively and sustainably, water systems and services need to be embedded within the social and geographical circumstances that pertain in specific communities. In other words, in order to be effective and sustainable, water systems and services need to accord with the cultural as well as the environmental particularities of place. Water is a cultural product in the sense that different people possess different knowledges of water, ascribe different meanings to water, and appraise and engage with water in very different ways. For example, while the determination of "good water" in one community might be on the basis of a scientific water quality analysis, another community might base its assessment on variables such as taste and colour, or on the type of raw water source. Similarly, technologies of water supply and water treatment that have evolved within and are appropriate to one set of social and environmental circumstances might not be appropriate to another. The adoption of inappropriate water systems and services is unlikely to succeed in the long term, as is the imposition of foreign approaches to appraising the safety and acceptability of water for drinking and other purposes. Because water is a cultural product as well as a technological product, sustainable solutions to the water problems in First Nations communities need to respond to the cultural reality of the places for which they are intended.

Demanding the right to water in terms of community control of the hydrosocial process in these places would help to ensure that water services are integral to the political and cultural reality of the communities involved. At the same time, local engagement in the process of developing and maintaining appropriate water technologies would help to build local pride and capacities in these communities and thus allow for the possibility of social transformation.

The right to water as a rule of social equity

The traditional formulation of the human right to water fixes definite quantities of water and the individual human body in a particular kind of relation. In the previous section we considered how the right to water can be understood to produce a different kind of relation – between the water process and society – to yield an argument for participation in collective processes of water governance, or management of the water process. Here, we will consider the right to water as a relation between the water process and society so as to yield an argument for sharing the economic benefits gained from the productive capacity of water. The right to water thus produces a rule of social equity and redistribution of wealth, which is applicable in circumstances where private appropriation of the water process yields profits in a market economy.

In the context of the present discussion, redistribution of wealth generated from the productive capacity of water is not only an end in itself but may be considered a means of realizing the traditional human right to quantities of clean water for personal and domestic use. This follows from the observation that even in places where such a right has been proclaimed in law – as in South Africa – poverty has prevented many people from realizing this right. As such, the question of access to water should be understood as a function of social and economic circumstances rather than a matter of technological fixes, rhetorical statements and solemn pledges, including the proclamation of the individual human right to quantities of water (Loftus, 2009, p953).

The idea of a right to water for such purposes as growing food and maintaining a healthy environment has been introduced above. While it is possible to quantify the water needed for such purposes, it is by virtue of water as a process that food production (growth/metabolism) and ecosystem functions take place. This broader notion of the human right to water was intimated in General Comment 15: While stressing that "priority in the allocation of water must be given to the right to water for personal and domestic uses", Article 6 of the General Comment affirms:

> Water is required for a range of different purposes, besides personal and domestic uses, to realize many of the Covenant rights. For instance, water is necessary to produce food (right to adequate food) and ensure environmental hygiene (right to health). Water is essential for securing livelihoods (right to gain a living by work) and enjoying certain cultural practices (right to take part in cultural life) . . .
>
> (United Nations Economic and Social Council, 2003, p3)

These observations underscore the contribution of the water process to the production process generally. If we consider the human right to water in terms of water's contribution to the production of goods and services, a different set of rights claims can be made upon the state from the usual claims that flow from the right to water for personal and domestic use. The human right to water then amounts to a social claim on the value of production. We are considering water here as something more than merely a factor of production but as a process that enables social production and reproduction, a sine qua non of the production of wealth and value. Water, as Amita Baviskar points out, is "an intrinsic element in the processes sustaining social production and reproduction." (Baviskar, 2007, p1) This function of water is reflected in the popularity of the concept of "virtual water", which refers to the volume of water embedded in the production of goods and services. Developed in the 1990s as a way of analyzing the virtual water content of foodstuffs traded internationally (Allan, 2003), the concept has since been expanded and applied to a wide range of goods and services (e.g. Hoekstra and Chapagain, 2008, Ch. 2). While the concept of virtual water helps reveal water's ingredient to the production process, it artificially fixes the water process in terms of specific quantities attached to classes of commodities. What we seek to do here is to valorize the hydrosocial processes that get fixed in the form of commodities for exchange.

Positing a right to water as intrinsic to the production process would have the effect of reclaiming for society the value of a process that has effectively been appropriated by private interests. Setting water to the production of commodities in a capitalist economy may be considered a form of what Marx termed "primitive accumulation" – "the historical process of divorcing the producer from the means of production." (Marx, 1978 [1867], p432) Where Marx was describing the early development of capitalism, others have pointed out that the basic process of appropriating resources that were formerly held in common in order to produce commodities for exchange has persisted through all phases of the expansion of capitalism. David Harvey identifies the current neoliberal variant of this process in terms of "accumulation by dispossession", noting that "many formerly common property resources, such as water, have been privatized . . . and brought within the capitalist logic of accumulation . . ." (Harvey, 2003, pp145–146) The process by which water is brought within "the capitalist logic of accumulation" has been further elaborated by Erik Swyngedouw: "Dispossessing H_2O", as Swyngedouw puts it (Swyngedouw, 2005) is an operation that occurs whenever water is withdrawn from nature and set to the production of commodities:

> The new accumulation strategies through water privatization imply a process through which nature's goods become integrated into global circuits of capital; local common goods are expropriated, transferred to the private sector and inserted in global money and capital flows, stock market assets, and portfolio holdings. A local/global choreography is forged that is predicated upon mobilizing local H_2O, turning it into money, and inserting this within transnational flows of circulating capital.
>
> (Swyngedouw, 2005, p87)

The idea of reclaiming a measure of the productive capacity of water as a human right may thus be understood as a matter of rightful repossession. Arguably, such a right is within the scope of emerging human rights discourse. Jurists have identified three different "generations" of human rights – the first comprising the classic rights of the individual and the second expressing collective rights such as in the right to work, leisure, education, and social security as set out in Articles 22–27 of the Universal Declaration of Human Rights. The third generation of human rights comprise what the Czech jurist and first Secretary General of the International Institute for Human Rights, Karel Vasak, describes as "rights of solidarity". "Such rights", Vasak notes, "include the right to development, the right to a healthy and ecologically balanced environment, the right to peace, and the right to ownership of the common heritage of mankind." (Vasak, 1977, p29)

It is acknowledged that these "third generation" rights are presently "more aspirational than justiciable in character, enjoying as yet an ambiguous judicial status as international human rights norms." (*Encyclopaedia Britannica*, 15th ed. vol 20, p717) Nevertheless, even an aspirational human right that recognizes "the right to ownership of the common heritage of mankind" suggests the possibility of beginning to redress a history of primitive accumulation – including the "dispossession" of the water process – within the framework of human rights discourse.

As for the practical means of realizing such a right, this may be built on the practice of having the state charge rents to commercial interests for the privilege of using water in their production processes and redistributing the value of these rents to the public. In Canada for example, the provincial state is generally considered to be the owner of water resources and has the constitutional authority to grant licenses for water withdrawals to industry and commercial enterprises. Each Canadian province has established a regime for granting permits for such purposes. At present however, the fees charged for these permits are considered to be very low – and in the case of several provinces they are negligible (Canada – National Round Table on the Environment and the Economy, 2010, pp49–51). The low rents charged for commercial water use in Canada are often criticized as economically inefficient in their failure to reflect the opportunity cost of water use, and for their failure to capture the value of ecological services provided by the water process (Renzetti, 2007). However, as resource economists Diane Dupont and Steven Renzetti have pointed out, by charging such low rents, the state misses out on the opportunity to redistribute a measure of the value that accrues to the private interests commanding the use of water in the production process:

> [B]y allowing industry, public utilities and farming operations free access to water resources, a provincial government is undertaking an implicit and poorly understood redistribution of wealth. This is because, by failing to capture some of the economic value created by the application of water in production processes, a government de facto allows that value to be directed toward those individuals and groups which have gained access to the use of fresh water resources . . .
> (Dupont and Renzetti, 1999, p365)

Renzetti argues further that "as the owner of a scarce and productive natural resource, the Crown [i.e. the provincial state] is entitled to share in the economic value created by the application of water in industrial processes ... Thus, it would seem fair to expect users to pay a reasonable fee in order to secure the use of a natural resource that contributes to their profitability." (Renzetti, 2007, p273) If we add the stipulation that the Crown redistribute the "economic value created by the application of water in industrial processes" for the general welfare of society, we have what amounts to the basic rule of social equity that is being put forward in this section. Formulating this rule in the idiom of a "third generation" human right as discussed above would lend greater force to its appeal. In this example from Canada, the realization of such a right is fully within the scope of existing legal authority and regulatory practice. It is suggested that similar authorities and practices could be considered as instruments for realizing such a right in other countries.

Conclusion

Considering the human right to water as constituting a kind of relation, this chapter has shown how such a right may be formulated in ways that go beyond the usual claim of a quantity of water for individual human needs. The right to water can thus be understood to define a relation between the collective identity of people on one hand and the process by which water articulates with society on the other. Specifically, two rights claims that flow from this understanding have been discussed: First, the right to water entails the right of participation in collective processes of water governance. Claiming such a right is conducive to realizing what have been described in terms of "community economies of water" (Bakker, 2007, p444, and chapter in this book) and "emancipatory water politics" (Swyngedouw, 2004, p177) so as to provide alternatives to neoliberalization. Second, the right to water entails the right of the collective to a share of the value generated in the hydrosocial production process. In addition to responding to the growing specter of the "dispossession" of water by private interests in the 21st century, both these expanded rights claims are conducive to a relational and dialectical process by which society changes itself through its collective engagements with water.

References

Allan, J. A. T. (2003) "Virtual Water – The Water, Food, and Trade Nexus: Useful Concept or Misleading Metaphor?", *Water International* vol 21, no 1, pp4–11

Assembly of First Nations (2010) "News Release, May 27, 2010: 'AFN National Chief Calls for Real Action on Safe Drinking Water for First Nations: Need

Action to Address the Capacity Gap as well as the Regulatory Gap' ", http://www. afn.ca/article.asp?id=4920, accessed 29 November 2010

Bakker, K. (2007) "The 'Commons' Versus the 'Commodity': Alter-globalization, Anti-privatization and the Human Right to Water in the Global South", *Antipode* vol 39, no 3, pp430–455

Bakker, K. (2010) *Privatizing Water: Governance Failure and the World's Urban Water Crisis*, Cornell University Press, Ithaca

Baviskar, A. (2007) "Introduction: Waterscapes: The Cultural Politics of a Natural Resource", in A. Baviskar (ed) *Waterscapes: The Cultural Politics of a Natural Resource*, Permanent Black, Ranikhet, India

Biswas, A. K. (2007) "Water as a Human Right in the MENA Region: Challenges and Opportunities", *Water Resources Development* vol 23, no 2, pp209–225

Bloomquist, W. and E. Schlager (2005) "Political Pitfalls of Integrated Watershed Management", *Society and Natural Resources* vol 18, no 2, pp101–117

Bowe, R. (2009) "The Human Right to Water – Corporate conference conveys concern, but activists decry the exploitation of dwindling fresh water supplies", in *San Francisco Bay Guardian Online*, http://www.sfbg.com/2009/12/16/human-right-water, accessed 29 November 2010

Brooks, D. (2007) "Human Rights to Water in North Africa and the Middle East: What is New and What is Not; What is Important and What is Not", *Water Resources Development* vol 23, no 2, pp227–241

Brown, J. C. and M. Purcell (2005) "There's Nothing Inherent about Scale: Political Ecology, the Local Trap, and the Politics of Development in the Brazilian Amazon", *Geoforum* vol 36, no 5, pp607–624

Canada – National Round Table on the Environment and the Economy (2010) *Changing Currents: Water Sustainability and the Future of Canada's Natural Resource Sectors*, National Round Table on the Environment and the Economy, Ottawa

deLoë, R. and R. Kreutzwiser (2007) "Challenging the Status Quo: The Evolution of Water Governance in Canada", in K. Bakker (ed) *Eau Canada: The Future of Canada's Water*, UBC Press, Vancouver and Toronto

Department of Indian Affairs and Northern Development (2006) "Report of the Expert Panel on Safe Drinking Water for First Nations (Volume 1)", Minister of Public Works and Government Services Canada, Ottawa

Dupont, D. and S. Renzetti (1999) "An Assessment of the Impact of a Provincial Water Charge", *Canadian Public Policy* vol 25, no 3, pp361–378

Gleick, P. H. (1996) "Basic Water Requirements for Human Activities: Meeting Basic Needs", *Water International* vol 21, pp83–92

Gleick, P. H. (1998) "The Human Right to Water", *Water Policy* vol 1, no 5, pp487–503

Hamlin, C. (2010) "The Philosophy of Water: The Gift of Good Water", unpublished paper presented at the University of Manchester, June 2, 2010

Harvey, D. (2003) *The New Imperialism*, Oxford University Press, Oxford

Harvey, D. (2008) "The Right to the City", *New Left Review* vol 53, pp23–40

Hoekstra, A. Y. and A. K. Chapagain (2008) *Globalization of Water: Sharing the Planet's Freshwater Resources*, Blackwell Publishing, Malden, MA and Oxford, UK

Irujo, A. E. (2007) "The Right to Water", *Water Resources Development* vol 23, no 2, pp267–283

Linton, J. (2010) *What is Water? The History of a Modern Abstraction*, UBC Press, Vancouver

Loftus, A. (2009) "Rethinking Political Ecologies of Water", *Third World Quarterly* vol 30, no 5, pp953–968

Marx, K. (1978 [1843]) "On the Jewish Question", in R. C. Tucker (ed) *The Marx-Engels Reader*, WW Norton, New York

Marx, K. (1978 [1867]) "Capital, Volume 1", in R. C. Tucker (ed) *The Marx-Engels Reader*, WW Norton, New York

Molle, F. (2009) "River-basin Planning and Management: The Social Life of a Concept", *Geoforum* vol 40, pp484–494

Norman, E. S. and K. Bakker (2009) "Transgressing Scales: Transboundary Water Governance across the Canada – U.S. Border", *Annals of the Association of American Geographers* vol 99, no 1, pp99–117

Nowlan, L. and K. Bakker (2010) "Practising Shared Water Governance in Canada: A Primer", UBC Program on Water Governance, Vancouver, http://www.watergovernance.ca/wp-content/uploads/2010/08/PractisingSharedWaterGovernance Primer_final1.pdf, accessed 29 November 2010

Phare, M.-A. S. (2009) *Denying the Source: The Crisis of First Nations Water Rights*, Rocky Mountain Books, Surrey, BC

Pradhan, R. and R. Meinzen-Dick (2010) "Which Rights are Right? Water Rights, Culture and Underlying Values", in P. G. Brown and J. J. Schmidt (eds) *Water Ethics: Foundational Readings for Students and Professionals*, Island Press, Washington

Renzetti, S. (2007) "Are the Prices Right? Balancing, Efficiency, Equity, and Sustainability in Water Pricing", in K. Bakker (ed) *Eau Canada: The Future of Canada's Water*, UBC Press, Vancouver and Toronto

Safe Drinking Water Foundation (2006) "Safe Drinking Water Foundation: Overview", http://www.safewater.org/PDFS/newsletters/Newsletter_SDWF_06.pdf, accessed 29 November 2010

Smith, N. (1992) "Geography, Difference and the Politics of Scale", in J. Doherty, E. Graham and M. Malek (eds) *Postmodernism and the Social Sciences*, Macmillan, London

Swyngedouw, E. (1997) "Neither Global nor Local: 'Glocalization' and the Politics of Scale", in K. Cox (ed) *Spaces of Globalization: Reasserting the Power of the Local*, Guilford Press, London

Swyngedouw, E. (2004) *Social Power and the Urbanization of Water: Flows of Power*, Oxford University Press, Oxford

Swyngedouw, E. (2005) "Dispossessing H_2O: The Contested Terrain of Water Privatization", *Capitalism, Nature, Socialism* vol 16, no 1, pp81–98

United Nations Department of Economic and Social Affairs (1992) "Agenda 21: The United Nations Programme of Action from Rio", UN Department of Economic and Social Affairs – Division for Sustainable Development

United Nations Economic and Social Council (2003) "General Comment No. 15 (2002) The right to water (arts. 11 and 12 of the International Covenant on Economic, Social and Cultural Rights). E/C.12/2002/11", Committee on Economic, Social and Cultural Rights

United Nations General Assembly – Human Rights Council (2010) "Human rights and access to safe drinking water and sanitation. A/HRC/15/L.14. 24 September, 2010."

United Nations General Assembly (2010) "The human right to water and sanitation. A/RES/64/292. 3 August, 2010."

Vasak, K. (1977) "Human Rights: A Thirty-Year Struggle: The Sustained Efforts to give Force of Law to the Universal Declaration of Human Rights", *UNESCO Courier* vol 30, no 11, pp29–32

4

A RIGHT TO WATER?

Geographico-legal perspectives

Chad Staddon, Thomas Appleby and Evadne Grant

Towards an enforceable right to water

One of the most vexed questions haunting the global environmental movement is the question of environmental "rights". We are said, variously, to have rights to a clean environment, to sufficient resources to support biophysical life and even socioeconomic development, to have a right to know about environmental quality, etc. Even the natural environment itself is sometimes said to have rights of its own, though how these may be articulated is a matter of considerable debate (cf Varner, 2002; Regan, 2004). However framed, one of the central rights often posited is the right to water. And this right does seem intuitively correct since, after all, water is absolutely central to human existence and to the human imagination. Biologically we are more than 2/3 composed of water and perhaps unsurprisingly, many cultures, around the world and down through the ages, have developed a complex and exalted conception of this simple but versatile molecule, two parts Hydrogen and one part Oxygen. H_2O. Some cultures even equate water with life itself (Staddon, 2010). Even a minor deficiency in water – say 5% of biophysical need – can seriously debilitate a human being. We can survive weeks without food, but only days without water (less in hotter, more arid places). Only air is more immediately necessary to survival.[1] And yet, as the economist Adam Smith pointed out long ago, there is the enduring paradox that while water, which is vital for life, is often considered valueless – a free good with which one can do what one will (though this is changing rapidly) – diamonds, which are biophysically useless, are highly valued. We may not like the direction of travel established by Smith and subsequent neoliberals who have sought to value water in very particular (market-dominated) ways, but we can surely recognize that these are strange priorities indeed! Be that as it may, there seem to be good reasons, prima facie, to propose not just a general recognition of a right to water, but its codification in laws, treaties, and binding conventions of any and every sort.

As recently as July 28, 2010 the UN General Assembly voted 122-0 in favour of a resolution declaring "the right to safe and clean drinking water and sanitation as a universal human right".[2]

Debating the right to water (in the UN General Assembly and elsewhere) might not be so pressing an issue were it not for the fact that so many people around the world currently do not have enough water to meet even their most basic needs. A significant proportion of the world's population already lacks even minimum quantities of clean water (defined by the UNDP as 50 litres per person per day (lpd) for drinking, cooking, cleaning and sanitation – cf. Gleick, 1999) not to mention the needs of industry and agriculture. While almost 100% of North Americans and Europeans have access to abundant clean drinking water in the home, only 28% of Kenyans, only 38% of Congolese, and only 69% of Mexicans have such access (Gardiner-Outlaw and Engleman, 1997; Gleick, 1999). The most recently completed global assessment, published in 2000 by the World Health Organization (WHO 2000) suggested that 1.1 billion people around the world lacked access to "improved water supply" and more than 2.4 billion lacked access to "improved sanitation". According to some estimates up to 2.5 million people die each year as a result of lack of clean drinking water and adequate sanitation services – that's 6800 people per day or 5 people per minute.

We further note that even in the so-called "developed world", access to water is becoming somewhat more asymmetric. In 2009, for the very first time, the British government recognized "water poverty" (defined as obtaining where households spend more than 3% of total household income on water services) as a serious social – and political – issue. Moreover many Western cities including Barcelona, Spain and Atlanta, USA have recently come close to running dry completely and some regions, such as central France, southeastern England and much of southeastern Europe face recurrent shortages (Staddon, 2010). Whilst there has been much attention to climate change as a key driver of growing shortages in absolute water supplies, the growing asymmetries are more directly the product of socioeconomic inequalities and a prevailing ideology that water is not "special"; that it is a market good just like fizzy drinks and holidays to Majorca.

So the prima facie case for a globally-articulated and globally-recognized right to water seems obvious and unimpeachable. There are some real problems of definition here though. For example just *who* is meant to be asserting *what sort of "right"* over *what sort of "water"*? Above we have been interpreting the question in terms particularly of the putative rights of people to an adequate provision of water and sanitation services. But these rights are not, as we shall see, to water itself, but rather to the benefits that access to it in certain qualitative and quantitative guises can bring, including the benefits of: hydration, hygiene, exercise and health, freedom from a polluted environment, and the right to know about the quality of the water environment. Inter alia all of these things can be asserted, but they are not about rights to water *itself*. Indeed, the right to water would seem to be, by itself, entirely meaningless. A more sophisticated understanding of the necessity of a right to water services ultimately depends upon understanding water not just as "drinking water" or "sanitation", but in all the many guises it takes throughout what we call, following Swyngedouw (2004) and others, the "hydro-social cycle": drinking water, atmospheric humidity, rain, a means of transport, something to paddle about in, an element of land-scape aesthetics, etc. And for water to do all these things we need to render our thinking about it much more sophisticated. Like Borges' famous Chinese Encyclopaedia, water is not water and by extension the juridical subjects meant to be possessing it are not necessarily who we think they are. It is therefore of critical importance to address the deeper questions embedded in the idea of a "right to water" – *who* is claimed to have *what sorts* of rights to *what sorts* of water-related services or benefits? Moreover, and from a legal perspective possibly even more importantly, *how* and *where* can these rights be enforced?

In addressing these last two questions we adopt a "geographico-legal" perspective. This hybrid perspective derives not just from the fact that the authors are by training an economic geographer (Staddon), property lawyer (Appleby) and a specialist in administrative law (Grant) but also from their joint recognition that legal doctrines, claims and expressions exhibit areal variations just as any other human phenomena. This is so not just because different localities (jurisdictions) have different laws and legal traditions, but because, to paraphrase Marx, law both makes space and is made by (and through) space. As argued by Chouinard (1994), Blomley (1994) and others there is geography *in* law as well as a geography *of* law. The relations between these twin geographico-legal realities are of particular importance for our analysis of the idea of a universal right to water.

Key legal expressions of the right to water

It is well known that the right to water has been expressed, if not precisely defined and enacted, in a series of national and international documents. Famously, and perhaps a first of its kind in the world, the 1996 Constitution of the Republic of South Africa (RSA) recognizes a specific right to water as follows:

27(1): Everyone has the right to have access to:

- health care services, including reproductive health care;
- sufficient food and water; and
- social security, including, if they are unable to support themselves and their dependents, appropriate social assistance.

27(2): The state must take reasonable legislative and other measures, within its available resources, to achieve the progressive realisation of each of these rights.

Although there is no quantitative guarantee in the South African Bill of Rights, the South African High Court, in the so-called "Phiri Case", ruled that the provision was meaningless unless linked to a 50 lpd level of provision, based on minimum levels of provision analyses by the World Health Organization (2000), Gleick (1999) and others.[3] Nonetheless, and as discussed in more detail in other chapters, in October 2009 the South African Constitutional Court (*Mazibuko v City of Johannesburg* Case CCT 39/09, 2010 (3) BCLR (CC), 8 October 2009) overturned the lower court's verdict, arguing that the general right enshrined in the 1996 Constitution does not preclude the imposition of limitations reflecting local conditions and constraints. In other words, Johannesburg is only obliged to provide water services "to the extent that it is reasonable to do so, having regard to its available resources".[4] While many commentators have been critical of the Constitutional Court's failure to endorse a quantitative guarantee, it is arguable that the decision rightly recognizes the need for careful consideration of both human needs and sustainability (Kotzé, 2010).

Of course expressions of and references to the right to water can also be found in a plethora of international human rights documents, including treaties, declarations and various sources of soft law. However, there remains considerable uncertainty regarding the legal status of the right. Cheers went up recently when the UN General Assembly (UNGA) adopted a resolution declaring "the right to safe and clean drinking water and sanitation as a human right that is essential for the full enjoyment of life and all human rights" (UN Doc A/64/L.63/Rev.1). However, it must be remembered that UNGA resolutions do not give rise to binding legal obligations in International Law. Although the resolution is an important indicator of international sentiment, it does not create an enforceable universal human right to water.

There are a number of international treaties that explicitly recognize the right to water, including the Convention on the Elimination of All Forms of Discrimination against Women (CEDAW), the Convention on the Rights of the Child (CRC) and Convention on the Rights of Persons with Disabilities. Although treaties do create binding obligations for those states who are party to them, in these instruments, the right is recognized in relation to specific groups – women, children, persons with disabilities – and it has therefore been argued that they do not provide an adequate legal basis for a *universal* right.

The International Covenant on Civil and Political Rights (ICCPR) and the International Covenant on Economic, Social and Cultural Rights (ICESCR) are the two most significant international treaties establishing universal human rights. However, neither of them make explicit reference to the right to water. Nonetheless it is often argued that since water is essential for life, the right to water is implicit in the right to life, recognized under article 6 of the ICCPR. Moreover, it is widely accepted that the right to water is essential for and therefore implicit in the right to an adequate standard of living, food and housing articulated in article 11 of the ICESCR and the right to health articulated in article 12. This view has been endorsed by the Committee on Economic, Social and Cultural Rights (the body responsible for interpreting and monitoring implementation of the ICESCR) in General Comment 15 on the Right to Water (26 November 2002, UN Doc E/C.12/2002/11). Although General Comment 15 in itself does not establish a legally enforceable universal right, it provides substantial weight to the arguments for the recognition of such a right.

These rights are about water for the satisfaction of *human* needs and potentials – there is a much larger body of law and treaty-making with respect to other water-related rights, particularly navigation, fisheries and riparian rights going back historically much, much further (Staddon, 2010). To date something like 150 international treaties concerning water have been signed, with the vast majority being bilateral, that is to say treaties between only two (usually riparian) nations. Less than two dozen truly multilateral treaties which bear on the human right to water have been ratified and entered into force (cf. Conca, 2006; Caponera and Nanni, 2007; Staddon, 2010).

A fundamental problem involves the complex legal identities of "water as property" and in particular what to do about potential clashing water rights principles, such as riparianism ("I can use it/own it because it flows through my land"), prior use ("first in time, first in right"), highest and best use ("based on judgments about optimality"), and equitable utilization ("all competing uses should have an allocation in proportion to their 'need'"). International treaties, perhaps because of their early preoccupation with transportation rights, tend to articulate the first and fourth of these rights principles. Thus, Article IV of the 1966 *Helsinki Rules* provides that "Each basin State is entitled, within its territory, to a reasonable and equitable share in the beneficial uses of the waters of an international drainage basin" – an expression in International Environmental Law (IEL) of the "riparian" doctrine, albeit one that raises the contentious issue of "equitability". Equitability can be judged many ways including per capita gains, absolute gains (economic, political, social, etc.), proportionality, etc. Even more confusingly, there is growing body of IEL which holds that water bodies of specifically "international" concern should be managed "equitably" – a principle that has actually been used to argue for water access rights for geographically-disconnected (non-riparian) third parties. This line of argument leads towards considerations of what is now commonly known as "hydropolitics" – the hydrological dimension of contemporary international relations (Ohlsson, 1995). Though beyond the scope of this chapter we note that "rights language" runs the risk of playing into the hands of those nations or regions seeking a legal basis for their water abstractions notwithstanding the potential for those abstractions to impinge on other "human" or other rights – for example the position that Egypt has consistently taken on the demands from upstream Nile riparian nations (Uganda, Ethiopia, etc.) to a greater share of Nile River waters than allowed under the (colonial) 1959 *Nile River Treaty*.

But usually when we posit a "right to water" we are referring to water in its biophysical utility – drinking water and water for sanitation, and the owner of the right is said to be an individual or class of individuals (e.g. "citizens of South Africa" or "children" in the examples noted above). These rights to water imply that the right should be for a sufficient *quantity* and also sufficient *quality* of water to meet biophysical or social development needs, since sufficient quantity is meaningless and without effect if it is not also of sufficient quality. This is what is implied in, for example, the 1989 *Convention on the Rights of the Child* which not only tries to prohibit the grosser abuses of children (as soldiers, as forced labourers, etc.), but also seeks to establish internationally the principle that it is a legal responsibility of signatory states (as of September 2011 only Somalia and the USA have not ratified the 1989 Convention) to ensure that children have the environment and resources they need to develop:

> To combat disease and malnutrition, including within the framework of primary health care, through, inter alia, the application of readily available technology and through the provision of adequate nutritious foods and clean drinking-water, taking into consideration the dangers and risks of environmental pollution.
>
> (Article 24, 2(c))

And the minimum quantity specified is linked to judgments about the minimum necessary for hydration, personal hygiene and public health, assuming minimal other infrastructural provision. Thus although Peter Gleick states that 50 lpd is minimally satisficing, WHO and emergency relief organizations often claim that a figure half that size, 20 or 25 lpd, is sufficient for meeting basic survival needs, increasing to 50–100 lpd if sanitation needs are taken into account. The Committee on Economic, Social and Cultural Rights elaborates on what the right to water entails in its General Comment 15 in even stronger terms:

> The human right to water entitles everyone to sufficient, safe, acceptable, physically accessible and affordable water for personal and domestic uses. An adequate amount of safe water is necessary to prevent death from dehydration, to reduce the risk of water-related disease and to provide for consumption, cooking, personal and domestic hygienic requirements.

Article 12(c)(2) of General Comment 15 further states that:

> Water, and water facilities and services, must be affordable for all.
> The direct and indirect costs and charges associated with securing
> water must be affordable, and must not compromise or threaten the
> realization of other Covenant rights.

General Comment 15 places specific obligations on signatory states to ensure
that this right is not interfered with through, for example, the imposition
of water tariffs by private water providers which are beyond the ability of
users to pay (as famously happened in the Cochabamba water privatization
case and in many countries of post-communist Eastern Europe – see Olivera,
2004; Staddon, 2010). As we argue in the next section, lack of adequate
enforcement mechanisms is a key stumbling block in ensuring the establishment of a meaningful human right to water. Moreover we contend that the
lack of a consistent and common enforcement mechanism has helped to
create an emergent geography of rights to water.

What sense are we making of all of this in the EU? Well, the most
comprehensive statement was made in December 2000 when the European
Commission adopted the *Water Framework Directive* as an attempt to create
a unified approach to water management out of the plethora of specific
technical directives developed over the past 20 years. Unfortunately for those
convinced by the argument for a "right to water", the WFD contains at its
very heart the so-called "Dublin Principles" (1992) which conceive of water
and water services as *property* which is alienable and whose value ought to
be determined by commodity markets. In the case of the Dublin Convention,
it is telling that the discourse of water scarcity is so easily conjoined with the
discourse about property to constitute what seems on the face of it an insur-
mountable challenge to the idea of water as a commons (*Guiding principles,
The 1992 Dublin Statement on Water and Sustainable Development*):

- Principle 1: 'Fresh water is a finite and vulnerable resource, essential to
  sustain life, development and the environment'
- Principle 2: 'Water development and management should be based on
  a participatory approach, involving users, planners and policy-makers
  at all levels'
- Principle 3: 'Women play a central part in the provision, management
  and safeguarding of water'
- Principle 4: 'Water has an economic value in all its competing uses and
  should be recognized as an economic good'

Here is another key challenge for those who would argue in favour of a universal human right to water – how to countermand the hegemony of the "water
as property" paradigm with the "water as commons" paradigm in such a way

that we can move towards a realization of the goal of sufficient water for all. As we shall see in the next section it is entirely possible to argue, and many in fact *have* argued, that if water is a human right then the appropriate mechanism for realizing that right is to treat it as a market commodity which can then be allocated according to ability to pay, with the state or other third parties stepping in where an inability to pay exists. In this way, argue advocates of water services privatization, not only will needed investment be brought into the sector, but provision will be rendered economically efficient (Staddon, 2010).

The 2004 *Berlin Rules on the Use of Water*, conceived as an attempt at the collation of numerous diverse separate acts, treaties and conventions, promisingly states in Article 4, chapter 17 that:

1   Every individual has a right of access to sufficient, safe, acceptable, physically accessible, and affordable water to meet that individual's vital human needs.
2   States shall ensure the implementation of the right of access to water on a non-discriminatory basis.
3   States shall progressively realize the right of access to water by:

- Refraining from interfering directly or indirectly with the enjoyment of the right;
- Preventing third parties from interfering with the enjoyment of the right;
- Taking measures to facilitate individuals access to water, such as defining and enforcing appropriate legal rights of access to and use of water; and
- Providing water or the means for obtaining water when individuals are unable, through reasons beyond their control, to access water through their own efforts.

4   States shall monitor and review periodically, through a participatory and transparent process, the realization of the right of access to water.

However, whilst there is no specific language about commodification of water resources (unless it is implicit in the definition of "optimal" use elsewhere in the *Rules*) neither is there a clear outline of enforcement mechanisms. Potential complainants are simply referred to "a competent court or administrative authority of that State" (Chapter XII, Art. 70) or, failing that:

1   The States or international organizations involved shall agree to submit their dispute to an ad hoc or permanent arbitral tribunal, or to a competent international court;
2   Recourse to arbitration or litigation implies an undertaking by the States involved in the dispute to accept any resulting award or judgment as final and binding.

Since it is generally only failed or authoritarian states that do not provide access to water resources complainants are left in the contradictory position of having to complain to the very states which have denied them in the first place! From this perspective another remarkable feature of the *Mazibuko v Johannesburg* case emerges: that it was actually possible to appeal the case through three levels of legal adjudication within South Africa and to nearly succeed. This however is the end of the (legal) road for the *Mazibuko* complainants – as one South African water rights activist put it to the authors "now the fight must go back to the streets!"

So we come back to the July 28, 2010 vote where by a margin of 122-0 with 41 abstentions the UN General Assembly passed resolution 64/292 entitled "The Human Right to Water and Sanitation". Although some of the abstentions cited the desire not to pre-empt other work in this direction already underway within the UN,[5] the General Assembly opted nonetheless to express deep concern "that approximately 884 million people lack access to safe drinking water and that more than 2.6 billion do not have access to basic sanitation, and alarmed that approximately 1.5 million children under 5 years of age die and 443 million school days are lost each year as a result of water- and sanitation-related diseases." Further the General Assembly urged UN member states "to provide financial resources, capacity-building and technology transfer, through international assistance and cooperation, in particular to developing countries, in order to scale up efforts to provide safe, clean, accessible and affordable drinking water and sanitation for all." While welcomed by water rights activists around the world it is difficult to see in this statement either clear definitions of what sorts of rights to water are implied (other than to avoid the perils specified above) or mechanisms for their successful exercise.

A legal deconstruction of rights discourse

Even if there is still no universal declaration of a right to water which is both prescriptive enough to be meaningfully tested in an accessible court and linked to a legal mechanism for its prosecution we certainly have had a plethora of statements, treaties and conventions claiming to take us in that direction. However, from a legal perspective such a desire is fraught with complexity and even danger, particularly where statements in international law but up against common law precedents. Reading writers such as Garret Hardin, it is easy to fall into a trap of believing that private property and commons exist in entirely different universes. As is well known, Hardin's arguments were based on a straw man inasmuch as the English commons were not a free for all but were governed prior to the enclosures by a complex system of overlapping and interacting commoners' proprietary rights and responsibilities. The Enclosures thus marked a paradigmatic replacement of a complex system of interacting rights with a simple regime based on "fee simple" ownership. Whether there was equity in that replacement process is a different question and is more likely to depend on the careful examination of each individual *Enclosure Act*.[6] This much is fairly well-understood, but it does not exhaust the range of issues raised by treating environmental resources as commons. For example, rights over resources are sometimes expressed in terms of the alternative. There is for instance a public right to fish in the sea and public rivers in the US, the UK and much of the common law world. This is a right to fish owned and used by the public expressed in law as a "public common of piscary". Lawyers would therefore take issue with the principle that property rights and common rights exist in entirely different paradigms.

Moreover, there is an (unfortunate) confusion, even amongst lawyers, regarding the distinction between property rights (whether common or private) and human rights. In respect of human rights – as enshrined in the ICESCR, the European Convention on Human Rights – water equity cannot easily be achieved by a simple replacement of one set of rules by another. For example, what if we were to produce a new annex to the European Convention on Human Rights which granted all citizens a right to 50 lpd, as suggested by Gleick (1999) or WHO (2000) and specifically contested in *Mazibuko v Johannesburg*? Would this achieve the desired aim of water equity? Well the answer is a resounding "maybe", but there would be an awful lot of unintended consequences along the way, and it is these that trouble environmental and human rights lawyers.

In the first place human rights legislation has a strange habit, in the United Kingdom at least, of assisting the very rich as well as the very poor, corporations as well as communities. The poor often have access to government funded legal aid and the rich can cover the prohibitive cost of the litigation, leading to situations where cases which may have no abiding public interest drag on inconclusively through the UK court system for years. With the current Conservative-Liberal Democrat government in the UK reducing the amount of legal aid available it seems that the legal field is being increasingly ceded to the wealthy.

Nonetheless in the United Kingdom it would appear that the very poor already benefit from a de facto right to water via the prohibition against domestic disconnection contained in the 1999 *Water Industry Act*. And since it is also illegal to supply water not fit for drinking or to limit its flow artificially (so called "trickle flow" or via pre-payment meters, as used in South Africa and elsewhere), then one could argue that in England and Wales at least there exists an enforceable and practiced right to water for domestic purposes. This does not extend to industrial or agricultural water users who are regulated differently. Since 1999 water companies have ceased disconnection of non-payers (average water bills in the UK are currently about £350/year/household) and pursue such debt through county court judgments (CCJs), notations on debtors' creditworthiness files and other civil collection mechanisms. While these procedures can bite in other ways (e.g. access to finance, mortgages, etc.) disconnections are a thing of the past in the UK.

If an additional human right to water was overlain onto this structure who would benefit? It is possible that companies and individuals whose consistent supply was threatened by some act of the state would be able to use this additional right to their advantage. Great care needs to be taken in the drafting of any human rights legislation both in respect of the protection of human rights to water and property rights over water. Turning to human rights first, let us look again at the well-known provision of the South African Constitution which states that "Everyone has the right to have access to . . . sufficient food and water . . ." A potential legal issue lies with the precise definition of the term "everyone". Since the rather notorious (in legal circles) American case of *Santa Clara County v Southern Pacific Railroad Company*, 118 US 394 (1886)[7] it is generally accepted that corporations not only have a legal personality (they are "juridical persons") which gives them a right to sue under Tort and other law but they also have the right of equal protection under the law (under the 14th Amendment to the US Constitution). The term "everyone" in this context may therefore include businesses and corporations. It raises the question: what is "sufficient water" for an intensive industrial process being undertaken by a legally-registered corporation? Could it be that in some cases South African industries could argue in court, with reference to Section 27(1) and Common Law precedents,[8] that their "rights" to water have equal standing with those of individual citizens or communities? Proponents of human rights to water seeking to use legislation to drive through what they take to be a more equitable distribution of water, may therefore find themselves in a legal "cleft stick", forced to argue for priority between two legally-equivalent juridical subjects. It is likely that intensive water users will scrutinize the same law to protect their own interests, and these interests are more likely to have the financial resources to take robust and prolonged legal action. Moreover since many corporations are now multinational enterprises they can more easily shop around for the most convenient courts in which to bring their actions, unlike ordinary citizens who are much more constrained in this respect.[9]

The second trap for the unwary is the treatment of water as a *property* right (see also chapter by Mitchell). Under the 5th Amendment to the US Constitution and Article 1 of the First Protocol of the European Convention on Human Rights an individual's possessions are protected from being taken by the state unless it is in the public interest and unless full compensation is paid. Again "individuals" in this context includes "corporations". In US jurisprudence particularly, regulating the use of property can amount to a "taking" (see for instance *Agins v City of Tiburon*, 447 U.S. 255 (1980)). Thus where a state or regulatory authority seeks to reduce existing water use it is possible to see a number of grounds for wealthy or corporate parties to use this aspect of rights law to protect their preexisting use. Note that water can also be a legal possession in a number of ways: there may be a long term abstraction right specifically granted by the state (often bundled together with land rights and allocations); there may be rights of abstraction given to riparian owners by the nation state's law; there may be rights of abstraction implied by longevity, historical priority, or by a state judgment about highest and best use. These rights then may take on the character of property and require full compensation if any alteration is proposed on any basis at all.[10]

From a legal perspective it is here that the dangers inherent in a human right to water are located. While the resource remains in the public trust with users exercising their legal rights to the resource by virtue of their being (equal) members of the public then the state is free to regulate it, without too much wrangling over compensation. The introduction of *individual* rights into this area would add an additional tier of complexity for regulators and the strong likelihood that corporations could (albeit unintentionally) benefit since the law applies equally to them as juridical subjects though they also have much greater resources to prosecute cases when and where they choose. For example, in the UK currently some water services companies have proposed, via the usual Coasean arguments about allocative efficiency, transforming the abstraction licensing regime in such a way that abstraction rights would become private property. Moreover, as we have seen above corporate water rights holders have been able to use property law to block state attempts at redistributing water for social or environmental gain, a global trend that greatly worries many water activists around the world (Barlow and Clarke, 2005). Surely this is an unintended consequence advocates to a human right to water would wish to avoid?

We need also to consider the "scale-effects" of a putative human right to water. In the United Kingdom domestic water distribution is controlled by a mixture of state entities (in Scotland, Wales and Northern Ireland) and privatized utility companies (England). All of these enterprises operate within a heavily regulated environment with a network of independent regulators controlling issues such as water and environmental quality, competition and pricing (Staddon, 2010). A domestic right to water via the European Convention on Human Rights, probably the most appropriate home for such a right in a European context, would necessarily overlay this existing regime, *and create in effect an additional regulator* via the European Court of Human Rights in Strasbourg. Such a court could, if it so chose, interpret the right to water more broadly than the narrow de facto definition currently operating in the UK (e.g. the right not to be disconnected) and in unpredictable ways. Furthermore the UK has already signed up to various general human rights statements and there is a question as to whether particular water rights would be needed in addition to already extant human rights (see also chapter in this book by van Rijswick and Keessen on the European Union).

In a developing world context the right to water is of course more of a pressing issue since it is in this context (rather than the European or

American contexts) that the need for better water services is most pressing. As noted above throughout much of the developing world, including many middle income countries (like Mexico), significant parts of the population do not have access to sufficient water for drinking or sanitation. Given the above discussion however we suggest that as much attention needs to be given to the *mechanisms* of monitoring and enforcement of a human right to water as is generally given in the literature to the substantive right itself. If a universal right to water were to be effective in the developing world, an additional regulatory burden on the developed world (in the form of UN-based monitoring and oversight) may, perhaps, be a reasonable "price to pay". This is particularly appropriate since adherence to the rule of law and access to justice can be highly problematic in some parts of the developing world, especially for marginalized members of society. Assuming these two features are present the question then arises as to how such a right will operate on the ground.

On the face of it, an international tariff set at 50 lpd would go some way toward resolving the issue of ensuring the satisfaction of any putative human right to water (as argued by the plaintiffs in the *Mazibuko* case). However, gaining international recognition of the right would not necessarily ensure enforcement of the right within states since international law (with the exception of the European Convention on Human Rights) is notoriously difficult to enforce especially by citizens against their own states. And even where the right to water is recognized within a national legal system, the judgment of the South African Constitutional Court in *Phiri* illustrates that what may be considered to be an internationally accepted norm (50 lpd) is not necessarily enforceable in a domestic context. And there is little the plaintiffs in *Mazibuko* can do about it – where after all could they take their human rights claim beyond the Constitutional Court, which has already ruled against them. And South Africa is not a "failed state" at all – imagine the considerably greater difficulties getting even as far as the plaintiffs in *Mazibuko* facing water-starved citizens in other countries with weaker and/or more corrupt states (see also chapters by Clark and Bond in this book).

On the other hand international law does have a persuasive capacity to spur national politicians and courts of most states – even weak ones – to take actions they might not otherwise take. It is possible that the inclusion of international legal obligations would encourage more countries to place onto their statute books legislation similar to the South African constitutional right to water which, although imperfect, would mark an advance for many of them. But such inclusions in international human rights law could ultimately be more important in influencing the behaviour of otherwise recalcitrant states. Moreover this could mark the expansion of what Bob Varady and others have called an emerging global "aperture of water governance" (Varady et al, 2003; Varady et al, 2009), particularly if the human right to water was linked somehow to the larger idea of "sustainability" which has a momentum of super-national IEL behind it:

> Ultimately, determining how best to achieve sustainability in multi-level water governance starts with a fundamental shift in the ontology of water governance: seeing GWIs as situated and operating within vital networks.
>
> (Varady et al, 2003, p.155)

International law is thus important for the development of human rights but the expansion of international jurisprudence is not confined to this area. International law is increasingly pervading all corners of daily life, via supranational organizations such as the European Union, this is in response to issues as wide-ranging as environmental degradation, to the law of warfare, and the regulation of international banks. So it is probably reasonable to argue that a right to water enshrined in international law would have some effect, even on recalcitrant or authoritarian states.

If it is accepted that there will be a long lead-in period, the question still remains as to whether a specific human right to water is required or whether it would be better to simply expand the existing human rights framework exemplified by such arrangements as the European Convention on Human Rights or the Conventions, Treaties, and Declarations noted above.[11] Consider the hypothetical position of a developing world village whose water supply runs short. In meeting its obligations does the relevant state conduct expensive capital works to provide water services *in situ*, or does it move the village to somewhere water is more abundant? Thinking about water rights *out of context* with other human rights can have unintended consequences. Both proposed solutions could theoretically meet a state's obligations in law, but the latter approach – moving the village – could be open to all sorts of abuses. In fact governments have worked this right typically the other way around – moving communities away from favoured large water development projects, as in the Three Gorges (China), Narmada Valley (India) and, most recently, the Giri III project in the highlands of Ethiopia.

Concluding comments: geographico-legal perspectives on the "human right to water"

The challenge posed by the idea of a human right to water may actually be usefully re-thought in terms of the need to re-insert an enforceable sense of the collectivity back into what Bob Varady and others have referred to as the emergent "global aperture of water governance". As we have seen above attempts to date have been rather halting and characterized by a distinct tendency to fixate on personal human rights to water, and on water rights overlooking the tendency for rights to manifest themselves in unexpected ways and places.

This then is the crux of the matter. The discourse of human rights may well serve as an instrument for the pathological expansion of neoliberal capitalism. But is this necessarily so, or only the tendency to date? This critique has been advanced by Brewster Kneen, Ziauddin Sardar and others who are specifically concerned that the postulation of rights to water necessarily presumes a specifically *neoliberal* political and economic subject. Though we consider this position to be overly-reductionist, we concur that this has certainly been part of the hegemonic legal discourse since at least the middle of the 19th century. What it is critical to challenge is the conflation of human rights with property rights, that is to say the fundamental Lockean solution to the problem of civil society (in the aftermath of the English Civil War and Restoration) of defining the core of modern political society in terms of our "ownership" of ourselves as a sort of non-alienable property.

So, having accepted the argument for a right to water, but having also due regard for the dangers inherent in its codification within an "enclosing" framework of the English Common Law tradition as well as the imperative that both water and its users be resituated within a framework of ontological equality, what can we now say to those immediately at risk of suffering due to a lack of sufficient clean water? By way of conclusion we offer the following points for consideration:

1. There is a large and complex body of already-existing international law codifying various sorts of rights to water in its many guises, including the satisfaction of biophysical need.
2. Most of this law-making however depends on signatory states to take the initiative, and it is often unclear what sorts of appeals aggrieved persons or collectives can make.
3. It is useful that International Law tends to subsume the right to water within broader outcomes-orientated obligations, as discussed above with respect to the UN Convention on the Rights of the Child.
4. Participatory forms of water management, such as those enacted through the EC Water Framework Directive (2000), offer a chance for alternative voices to be heard, and there are other non-European moves in this direction such as the Nile Basin and Mekong Basin Commissions.
5. Further, it may be possible in certain cases for individuals or non-state bodies to bring water rights cases before the European Court for Human Rights or the Inter-American Court for Human Rights, but one would have to be careful that the assertion of the tort was direct and not frivolous or vicarious – e.g. a claim by concerned English activists against Veolia for preventing poor African communities from getting water would probably fail to be heard, unless the deprivation of water to the African communities are incidental to a tort suffered by the Europe-based plaintiffs (for example if the English plaintiffs were shareholders).

Finally we would suggest that international water rights activists should be very careful what they wish for. Many formulations of the right to water can just as easily reinforce the rights of existing users, including corporations.

Notes

1 Strangely, Abraham Maslow, whose famous "hierarchy of needs" is well known throughout the social sciences, ranked the need for water *third*, right after "breathing" and "food"!

2 Forty-one nations, including the Netherlands, Canada and the USA abstained on a variety of grounds including the contentions that the resolution was not sufficiently robust and enforceable (the Netherlands), that it was somehow "premature" given the on-going work of a specially-commissioned UN rapporteur (Canada, the USA, the UK, Australia).

3 The World Health Organization recommends a minimum of 20 litres of water per person per day for basic survival, and 50 to 100 litres per day per person to meet most health needs, including public sanitation and hygiene.

4 Writing for a unanimous Court, Justice O'Regan held (1) the City's free basic water policy of 25 lpd to be reasonable under section 27(1) of the Constitution, and (2) the introduction of pre-paid water meters to be lawful, procedurally fair, and not unfairly discriminatory.

5 In late 2008 the Human Rights Council appointed Catarina de Albuquerque as the Independent Expert on human rights obligations related to safe drinking water and sanitation. She is due to due to report on her work to the Human Rights Council in October 2011.

6 For example, the 1668 Act of the English Parliament which enclosed the lower half of what is now Lower Woods Nature Reserve in South Gloucestershire did extinguish commoners' rights of longstanding including pannage, deadwood collection, etc. However, the Badminton Estate to which the lands in question passed at about this time neither enclosed the land nor enforced the extinguishment of commoners' rights which have continued through to the present day (see Burditt, 2008).

7 The notoriety stems from US Chief Justice's unorthodox decision to leave this aspect of the case for the Court Reporter to decide.

8 Non-lawyers need to understand that legal codes are always subject to interpretation, especially – in the Common Law states – with reference to previous judgments and precedents in like states.

9 Consider the use of UK courts by corporations from all over the world to bring libel and other actions.

10 In a more recent judgment the US Supreme Court has struck down the "legitimate state interests" test implied in the *Agins v. Tiburon* judgment without replacing it, further muddying the already murky legal waters.

11 In fact Article 2 of the *European Convention on Human Rights* ensures that "everyone's right to life shall be protected by law", and as water is an intrinsic part of life it is likely that water rights are *already protected* by the Convention.

References

Barlow, M.; Clarke, T. (2005) *Blue Gold: The Fight to Stop the Corporate Theft of the World's Water*, The New Press, New York

Blomley, N. (1994) *Law, Space and the Geographies of Power*, Guilford, New York

Burditt, T. (2008) "Field Working: exploring knowledge networks in the practice of nature conservation", unpublished PhD thesis, University of the West of England, Bristol, England

Caponera, D.A.; Nanni, M. (2007) *Principles of Water Law and Administration*, Taylor and Francis, London

Chouinard, V. (1994) "Geography, law and legal struggles: which ways ahead", *Progress in Human Geography*, vol. 11, no. 5, pp.415–440

Conca, K. (2006) *Governing Water: Contentious Transnational Politics and Global Institution Building*, MIT Press, Cambridge, MA

Gardiner-Outlaw, T.; Engleman, R. (1997) *Sustaining Water, Easing Scarcity: A Second Update*, Population Action International, Washington, DC

Gleick, P.H. (1999) "The human right to water", *Water Policy*, vol. 1, no. 5, pp.487–503

Kneen, B. (2009) *The Tyranny of Rights*, The Ram's Horn, Ottawa

Kotzé, L.J. (2010) "Phiri, the plight of the poor and the perils of climate change: time to rethink environmental and socio-economic rights in South Africa?", *Journal of Human Rights and the Environment*, vol. 1, pp.135–160

Ohlsson, L. (Ed.) (1995) *Hydropolitics: Conflicts over Water a Development Constraint*, Zed Books, London

Olivera, O. (2004) *Cochabamba! Water War in Bolivia*, South End Press, Cambridge, MA

Regan, T. (2004) *The Case for Animal Rights: Updated with a New Preface*, University of California Press, Los Angeles

Sardar, Z. (1998) *Postmodernism and the Other: The New Imperialism of Western Culture*, Pluto Books, New York

Staddon, C. (2010) *Managing Europe's Water Resources: 21st Century Challenges*, Ashgate Press, Farnham, Surrey

Swyngedouw, E. (2004) *Social Power and the Urbanization of Water – Flows of Power*, Oxford University Press, Oxford

Varady, R.; Meehan, K.; McGovern, E. (2009) "Charting the emergence of 'global water initiatives' in world water governance", *Physics and Chemistry of the Earth*, vol. 34, pp.150–155

Varady, R.; Meehan, K.; McGovern, E. (2003) Global water initiatives: some preliminary observations on their evolution and significance. Proceedings of the 3rd Conf. of Intl. Water History Assoc, Alexandria, Egypt, 25 pp

Varner, G.E. (2002) *In Nature's Interests? Interests, Animal Rights, and Environmental Ethics*, Oxford University Press, New York

World Health Organization (2000) *Global Water Supply and Sanitation Assessment 2000 Report*, World Health Organization and United Nations Children's Fund, New York

5

THE POLITICAL ECONOMY OF THE RIGHT TO WATER

Reinvigorating the question of property

Kyle R. Mitchell

"It would be difficult to argue that human rights prevail on the world stage and not the principle of might is right."

(Teeple, 2005, p1)

Introduction

The condition of constant and rapid expansion is the impetus for the astonishingly creative and productive nature of capitalism. This need for perpetual growth is also the driving force behind neoliberal mandated policy prescriptions that have taken shape throughout the world in the dismantling of government mandated regulatory frameworks of varied sorts (most notably environmental, health and labour) as well as the withdrawal of the state from the social provision of what is commonly known as social and/ or public goods and services, including water and wastewater services. Though these neoliberal reforms are extremely complex and diverse on a case by case basis – in both form and scope – and also highly variegated in terms of intensity of implementation, the overarching modus operandi of these reforms is the penetration, expansion and accumulation of capital into areas of life hitherto unaffected by the free market.

These policy prescriptions have been concretized in large part by way of the processes and mechanisms of economic globalization, yet recently they have been reignited, albeit repackaged, according to a purported need to cut public expenditures in the face of the recent global financial recession (2007 and continuing). In terms of overall "mental conceptions"[1] of publicly delivered goods and services, this repackaging of neoliberal policies by way of macroeconomic policy reform has predictably done for capital accumulation in the last year and a half what economic globalization has intended for and indeed largely accomplished in the last 35–40 years – that is, systematically call into question and incrementally dismantle, where and when possible, forms of social reproduction that exist outside that of a private property framework.

Although economic globalization and its impact on the water and waste-water sector has been well-documented and satisfactorily assessed by a substantial and diverse body of critical literature from various academic disciplines and activist work, much of this corpus deals with the principal policy objectives of its neoliberal character. Indeed, this literature has done much to raise awareness about the social, political, economic and cultural outcomes of such reform as well as to document and analyse the political persuasion and scope of the attending global governing institutions that encourage and in some cases outright implement such reform. What it often misses, however, is analysis that probes the prevailing property relations, defined by private property, which presuppose these aspects and outcomes of neoliberal reform; indeed, these relations make neoliberal reform a possibility in the first place. Just as these prevailing property relations ultimately frame the way in which production, distribution and consumption is carried out within the global economy, they are a product of and define the struggle between competing interests within liberal democracies. It is proposed here, that a return to the political and economic foundation of the global economy, with a robust analysis of its corresponding liberal democratic political expression and the attendant sphere of civil society, is necessary to fully comprehend the complexity of the struggle over the right to water.

To this end, the aim here is to provide a more nuanced reconceptualization of the struggle over "water rights" and in so doing to recast this contestation as a struggle over the *right to water* in the context of the unequal property relations that characterize the global economy. By way of a political economy approach then, this chapter proposes a theoretical perspective of the struggle over the right to water by reinvigorating the question of property in the context of the competing interests over the right to water. It follows that what is also required is a recalibration of the concept of civil society so as to fully comprehend the unequal power relations characterizing the contradictory interests involved in the struggle over this right. By conceiving of civil society as a terrain of contested relations (Keck and Sikkink, 1998) rather than a neutral space of free association and volunteerism, as it is so often thought, this chapter explores how the struggle over the right to water is fundamentally defined by the broader unequal property relations that characterize the capitalist mode of production and its political expression liberal democracy. By way of conclusion, this chapter calls into question struggles that limit their contestation of the right to water within the circumscription of the liberal democratic framework.

The question of property in relation to the right to water

The rights discussion is instigated by a series of complex separations, namely, the separation of ordinary individuals from: their communities (social); the political process (political); and direct access to the means of subsistence and therefore the ability to carry out and sustain a healthy way of life – this particular separation is presupposed by the separation of the exchange value of a good or service from its use-value (economic) (Brown, 2009). As these relations relate to the question of the right to water, they occur on corresponding levels. In this way, ordinary individuals are separated from: responsibility (not to mention moral obligation) to see that others' water needs are met (social); the decision-making processes related to the future of available water (political) and; direct access to water for cleaning, cooking, farming and drinking, by way of a price mechanism (economic).

Either considered in conjunction or as isolated albeit related processes, these separations are often described historically as the enclosure of the commons. Although enclosure takes many different forms, spans centuries and reaches all corners of the globe, taken together, these processes represent a transformation in social relations, from collective or communal property relations to private property. In one way or another, this transformation, or what has been typified by Harvey (2003) as "accumulation by dispossession" – facilitated today by economic globalization and its neoliberal political expression – represents not just a modern example of what was once referred to as "primitive accumulation", but represents rather, the continuation of capitalist bound primitive accumulation. In other words, the outcomes (increase commercialization) and consequence (increase material inequality) of the political and economic processes that transform the character of public goods and services, such as water and wastewater services, are the logical and essential trajectory of the capitalist system. Yet how are we to understand this historical process in the context of water and wastewater services? This question is inspired, in part, by Brown's argument that in addition to the abovementioned separations there is a necessary ideological separation that obscures the class character of separation and thus "any sense of historical causality", which, in turn, provides capitalism the necessary stability to expand during times of contentious struggle (Brown, 2009, p576). We turn to the question of property in order to tease out the relational qualities of rights in an attempt to seek greater understanding of the struggle over the right to water.

The concept of property is central to the political economy of water. From the outset, some may interpret the suggestion to reinvigorate the question of property in regards to water as the objectification of social relations and the biophysical world – an attempt to place fluid processes into fixed categories or measurements. Property in this sense refers to a tangible thing, a resource unit or a physical object such as a good or service. Indeed, it is common to think of property in this way and so too is it ever-increasingly commonplace with water. This common conception of property, many argue, originated as a result of the social division of labour, or, in other words, when people began producing things or providing services strictly for exchange instead of immediate and direct consumption (MacPherson, 1978).[2]

The exchange of goods and services, however, was one of humanity's "first historical acts" as it was famously stated and should not be confused with the ascendancy of the free market (e.g. tribal exchange was based on barter).[3] Furthermore, an uncritical acceptance of an ahistorical perception of the free market obscures the exploitative and often violent conditions that characterize the enclosure processes that lead to the creation of markets of one sort or another (e.g. water market). Because of the pervasiveness of the capitalist relations of production, however, and the inevitability and prevailing propensity towards commodity production and exchange pursuant to these relations (those based on price to which money is the value form) people have come to associate property with inherently possessing exchange value, instead of use-value (MacPherson, 1978). Yet, the view that property is a tangible object or commodity is entirely correct and consistent with its dictionary definition; however, it misses a second sense of the definition – the legal sense – that separates the physical or mystical[4] (i.e. commodity form) from what is actually being possessed.

We argue here that a return to the second sense of property, and the reconstitution of the idea of property as a legal relation to the proposed political economy of water as a theoretical framework, serves as a robust measure upon which we may assess: first, the extent to which the enclosure of water exists; second, the extent to which these enclosures contradict humanity's shared reliance on water; and third, the extent to which resistance movements opposing the commodification and privatization of water structurally challenge the prevailing economic order and its corresponding trajectory and political expression.

The second sense of property, then, has to do with its legal significance. Legally, property is not a thing but instead a right. This distinction – property not as a physical possession but as a relationship – is central to the reassertion of property as a legal definition as well as, and most importantly, a social relation. Inasmuch as property is an abstraction so too then is any discussion of rights as rights separate the physical from what is actually possessed. Indeed, it has been said that property is best understood relationally. Arnold's (2002) "web of interests" metaphor for property is symbolic of the nature of property relations encapsulating the many interests, shared and/or competing, all of which are interconnected and involved in the question of rights. Rights most fundamentally confer an individual's relation to a good or service such as water. This relation, however, is predicated on a need to assert such rights – to stake claim to a good or service in the presence of others. In other words, rights to goods and services are predicated on and also determine our relations with other people. Bryan points out, "Property is an expression of social relationships because it organizes people with respect to each other and their material environment" (Bryan, 2000, p3). Property, then, is a right in the sense that people possess claims or entitlements to goods and services broadly defined. Property, in this sense, is synonymous with the meaning of right (MacPherson, 1978; Teeple, 2005).

What is actually being possessed, then, or what is in fact owned, are rights to goods and services broadly defined. This ownership can be exercised collectively or privately. Most importantly, any given system of property and the corresponding entitlements and claims associated with its rights point to the fact that any system of production and exchange define actions individuals take in relation to others (Heritier, 2002). That is to say that all property arrangements are fundamentally based on social relations and not as often otherwise suggested – the natural unravelling of human history. In this way, rights point to the social nature of all economic systems and social formations, thus directing our attention to the relational qualities of our material existence, including the production, distribution and consumption of socially necessary goods and services. Linton (in this book) offers a comprehensive discussion of the relational approach to the right to water therein supporting a rights structure that more appropriately reflects both humanity's shared reliance on water and the "processual" nature of water, thereby satisfying the "hydrosocial arrangements" constituting these relations.

Access to socially necessary goods and services, such as water and wastewater services, is based on legal (formal) or informal entitlements or claims. That is to say that both legal and informal entitlements and claims are the embodiment of socially sanctioned and therefore socially legitimated power – legal rights are formalized and ensconced in law and enforced by an institutional power such as the state (through government/legislature, military, police and security forces and the courts/law) and informal rights are maintained, reproduced and enforced by informal group customs and practices, expectations or norms (MacPherson, 1978; Anderson and Simmons, 1993).

Despite common perception, property is not synonymous with private property or exclusive individual rights. This association, albeit a misconception, has much to do with the fact that private property is the dominant form of property within the capitalist system. As already mentioned, the free market is often treated as self-evident, and, as such, property is not only assumed to be a commodity (therefore susceptible to the first sense of the definition) but it is also associated with the exclusive individual rights, such as how the human right to water is conceived: as an individual right. This misconception is regarded by many critics of neoclassical economics as a deliberate misconstrual with the intention of perpetuating the idea that private property is superior to all other property arrangements (perpetuated by the "tragedy of the commons" metaphor) except in those cases where there has been proven market failures or improper governance.[5] Indeed, in many cases, property is possessed collectively and exercised as a collective or common right so that members of a particular community (be it local, tribal, state, etc.) share in the use of a common resource. A recent conceptual development within the right to water debate, characterized by a shift from the human rights-based language to that of the idea of the water commons, as noted by many authors in this book (notably Linton, Bond and Clark), reflects the nature of collective rights to water. Furthermore, as Staddon et al argue, this contextualization is compatible with conceptualizing water as a property relation; that is, conceiving of water relationally, as it is argued here.

Even still, private property is the prevailing property relation within liberal democracies. As private property relations constitute the hierarchical structure of the capitalist relations of production (Anderson and McChesney, 2003), including the economic (broader socio-economic structure) (Hinkelammert, 1986), political, legal and cultural spheres (Caruthers and Ariovich, 2004; Reeve, 1986), and as liberal democracy is the dominant political expression of this mode of production, it is here, in the constitutive frameworks of both the economic and political system, where we will centre our analysis of property. How, then, does this prevailing relation affect the struggle of the right to water? An analysis of the competing relations that make up civil society will bring us closer to a more comprehensive grasp as to how power relations affect the outcomes of the struggle over the right to water.

Liberal Democracy and civil society: a contested concept in a terrain of contested relations

In times of confusion, our will to understand the world blends naturally with our desire to feel at home within it. In such times, we sometimes struggle so hard to make sense of the world that we forget that the senses we make change this world in unexpected ways. And as a result, we frequently end up falling prey to the very same forces we so dearly want to escape.

(Bartelson, 2006, p371)

Though the idea of civil society has become ubiquitous with global governance and economic globalization, the character and objectives of so-called civil society are, at best, ambiguous (Kaldor, 2003; Amoore and Langley, 2004; Teeple, 2005; Bartelson, 2006). This uncertainty has much to do with the concept's historical development (Kaldor, 2003; Bartelson, 2006). Civil society has various time-bound expressions that have straddled equally divergent political and economic models, and each of these develop concomitantly with and indeed embody not only the contestation over what actually constitutes civil society (i.e. its constituent parts) but also the real competing relations it actually comprises. These factors have led to what Bartelson (2006) refers to as the incessant "ideologization" of the concept. What remains amidst all of the ambiguity and historical difference, however, is that civil society is still a contested concept (Kumar, 1993; Castiglione, 1994; Chandhoke,

2002; Keane, 2003; Chandler, 2004; Anheier, 2005; Teeple, 2005). Even so, it would markedly appear that the vivacious spirit of critical engagement that once characterized its analysis and constant reconfiguration has vanished from the scholastic purview, thus leaving an even greater yet unquestioned confusion over the very idea of civil society itself.

The idea of civil society was reignited and indeed re-articulated by the coinciding of two movements in the late 1980s and early 1990s. Oppositional movements in both Eastern Europe and Latin America were challenging authoritative governments in their respective countries while a movement of radicals across the globe were calling into question the undemocratic nature of a maturing global economy (Kaldor, 2000; Chandler, 2004; Bartelson, 2006). These movements were carving out a space for progressive politics to be known as global civil society. For the first time, single-issue groups linked their struggles against global capitalism and in doing so formed a consolidated albeit highly unorganized movement resisting the neoliberalization of public life and the biophysical world. It was around the time of the "Battle in Seattle" in particular that global civil society received contemporary currency amongst activists in the alter-globalization movement as well as academics and other commentators assessing the impacts of globalization. It is widely accepted that the water justice movement has matured concomitantly with this understanding of civil society. As the constituents of water justice organizations or as the mobilizing force behind progressive water justice demonstrations and gatherings, civil society is thought to be a somewhat coherent singular force promulgating water justice in an inequitable transnational political milieu. Rarely, however, is the notion of civil society critically examined. Yet, the question of civil society is of central importance to the struggle over water because it not only highlights where the water justice movement may situate its struggle but also how this movement defines struggle.

What is particularly noteworthy about the modern conception of civil society and that which transformed the constitutive character of its historical conception is that contemporarily civil society has been reinvented as a political space for dissenting voices exclusively (Cohen and Arato, 1992; Gellner, 1998; Kaldor, 2003; White, 1994; Walzer, 1995). Global civil society emerged as an open space largely for non-commercial entities and action. This includes individual activists, activist movements, non-profit organizations and non-governmental organizations alike. This contemporary conception would, as Chandhoke suggests, "seem to be supremely uncontaminated by either the power of states or that of markets" (Chandhoke, 2002, p36). In this way it came to be known as the third realm (Cohen and Arato, 1992) or the third sector (Chandhoke, 2002).

The mainstream avatar of civil society as a third sector or the social economy, generally views liberal democracy as constituting three main elements, the component parts of which in relation to global water conflict include: (i) the corporate sector, including companies specializing in the construction and/or provision of water and wastewater utilities and services as well as a transnational alliance of pro-privatization think tanks and global governing institutions; (ii) the water justice movement resisting privatization and commodification; and (iii) the state whose purpose is to maintain a balance of power between competing interests and deliver certain public goods and services upon receiving the time-bound social legitimacy to do so.

As evidenced from this simplistic yet conceptually precise avatar, so cumbersome and imprecise the concept of civil society has become that every use of it seemingly signifies any progressive actor, organization, institution or relation positioned outside the market sphere actively opposing neoliberalism (White, 1994; Walzer, 1995; Gellner, 1998). Yet widespread and contending uses of the concept – from anti-capitalist resistance mobilizations, to socialist or liberal democratic non-governmental organizations, to the capitalist global governing institutions such as the World Bank and others – illuminate the highly contentious and contradictory use of the concept thus calling into question civil society's constitutive character and elemental form, not to mention the context to which the idea itself is employed. Though the competing uses of civil society is indeed an area of social inquiry demanding much attention, it is only indirectly that we introduce this problematic so as to illustrate its contending use. Our immediate concern, rather, is with how the concept of civil society, and its critical engagement in particular, relate to an effective way in which to theoretically conceptualize the struggle over the right to water.

Critiquing this common conception of civil society is a delicate task to be sure. Bartelson concurs suggesting, "Opposing civil society on political ground is difficult, and risks stigmatizing the prospective opponent" (Bartelson, 2006, p387). Radical critique of its contemporary conception is indeed negligible; moreover such critique is largely missing in the context over the struggle over the right to water. Yet, the contemporary conception of civil society obscures the property relations that inform the historical trajectory of the concept, including the unequal material conditions that foster the ever-growing disparity between classes, those relations which have a direct bearing on the question of access to and control of water.

Socio-historical accounts of the idea of civil society, post eighteenth century,[6] unanimously document the underlying commonality amongst all hitherto conceptions, that is, the recognition of the divorce between civil society and the state. What is missing from many analyses of civil society is the question as to what provoked this separation. Most treated the evolution of a set of property relations conducive to commodity production and a division of labour, as the historical precondition for the emergence of civil society. These relations took root in a series of upheavals, prompted by the Age of Enlightenment, which were responsible for the proliferation of liberalism as a political and economic doctrine. Bellamy suggests, "historically, the most significant social influence on the formation of liberalism was the passage from feudalism to capitalism . . . (where) . . . an individual's social position and success supposedly mirrored his or her ability and effort – a way of life they felt was best realized in a free market economy" (Bellamy, 2001, p8798). In this way, social structure would no longer be based on status as it was under feudal relations, but rather by the contractual relations between individuals in the marketplace (Bellamy, 2001). This development further entrenched the idea that the individual, not the community, was at the centre of all social life (Teeple, 1984). These property relations fundamentally transformed the way in which goods and services were to be produced and exchanged.

The division of labour and the corresponding development of a class-based system would structure civil society relations and thenceforth social relations would be defined by contradictory or competing property relations. Conflict between individuals, as well as between individual interests and those of the broader public interest, would become contentious and therefore require a governance structure to mediate relations in an emerging market society. Such a framework would be institutionalized in the state and express itself by way of a corresponding political and economic philosophy – classical and economic liberalism – that would find its expression in society by way of the rule of law which ultimately is the expression of the balance of class power of any given historical epoch. Individuals were endowed with civil rights (mediating conflict between individuals) and political rights (mediating relationship between individuals and the state). These rights became the personification of abstracted and indeed estranged relations.

The political form corresponding with this separation would take form in liberal democracy. Liberal democracy emerged as a formalized framework of politics and governance in the late eighteenth century and would come to dominate advanced capitalist economies in the nineteenth century, as it does today (Bellamy, 2001). Liberal democracy, as a political expression of the capitalist mode of production, is not only the manifestation of the separation between civil society and the nation state, but also the embodiment of competing and often contradictory interests. As aggregated capital began running up against forms of individual capital (ultimately leading to the demise of petty-commodity production) the dominant form of the private property relation at the national level would become corporate private property. As Teeple notes, "the success of incorporated collective, rather than individual capital becomes the modus vivendi of the system" (Teeple, 2005, p36). The dispossession of ordinary people from their means of production developed a working class and this process would come to characterize the growth of advanced capitalism.

Corporate rights increasingly eclipse all other forms of rights particularly those that exist outside that of the private property framework (as these relations are seen as unproductive or not producing profit). In areas of life where class struggle forged the development of social rights such as the right to water and wastewater services, these concessionary property relations would not only protect ordinary people from the worst effects of the capitalist system but also maintain a relatively healthy labour force for the capitalist market. Social rights, as such however, can be viewed as contradicting the preeminent corporate private property relation and in this way they may be seen as countervailing. Inasmuch as they may "represent a balance between conflicting demands . . ." as Frow notes, they "are always contested" (Frow, 1996, p106). Legislation which comprises the provision, protection and/or extension of social rights, such as the case with publicly delivered water services, is only as robust and relevant as is the broad-based support for it. At no other time in history have institutionalized social rights been eclipsed by the preeminent right of the corporation than in this era of economic globalization where the contestation over collective rights has intensified. The separation of people from political power, embodied in the divorce between civil society and the nation state, becomes ever more plain to see in the transnational sphere where the nation-state's ability to make decisions in the benefit of ordinary people instead of global capital becomes circumscribed by global institutions and charters whose sole purpose is the expansion of the rights of capital globally (Teeple, 2000).

Where there is significant opposition to the expansion of capital in areas of daily life protected by social rights, including water and wastewater services, corporations use their disproportionate power – economic, political and social – to assert their interests. In places where public opinion is overwhelmingly at odds with corporate interests, then, the intensity of coercion and violence, if need be, is elevated. Take the 2000 uprising to water privatization in Cochabamba, Bolivia, as an example. Thousands mobilized in resistance to water austerity measures and as a result the government declared martial law which provoked violent riots between police and protesters. Only after the death of a protester and hundreds of injuries did the government call off the police and military.

Where violence is not necessarily needed, corporate power is exercised by way of lobbying and public relations campaigns, or what Miller and Dinan refer to as the "products of diligence, hard work, planning and conscious ideological warfare" (Miller and Dinan, 2008, p1). Corporations use these campaigns to direct concerted attacks on those areas of life yet to be directly penetrated by corporate capital and in so doing influence mental conceptions. Miller and Dinan prefer to use the idea of propaganda instead of public relations as it more appropriately typifies the entire set of actions taken by corporate interests in these contexts; in other words, propaganda "implies the unity of communication and action taken by corporations" (Miller and Dinan, 2008, p5). The contexts in which the expansion of capital is debated are deluged by what Miller and Dinan argue as "contemporary 'common sense' that what is good for business must be good for society" (Miller and

Dinan, 2008, p3). The struggle over the right to water in Scotland is a case in point as this example has become the most recent battleground for water justice.

A cursory appraisal of the historical trajectory of commercialization of water and wastewater services in Scotland replete with a coalition of business interests, neoliberal think tanks, media elements as well as stakeholders and regulators from the water and wastewater sector endorsing commercialization (Kane and Mitchell, 2010) serves to illustrate how unequal power relations compel macro-economic reform in favour of capitalist logic despite widespread resistance. Following the privatization of water services in England and Wales in 1989 it was the legislative intention of the ruling Conservative Government of the United Kingdom to bring Scotland in line with this agenda. The 1994 Strathclyde water referendum experienced a 74% turnout rate with 97% voting against privatization and sent an unequivocal message to the Westminster Parliament that the majority of voters were against commercialization. Following Scotland's devolution, however, and the merging of Scotland's three water authorities into one ever-increasing corporatized entity, Scottish Water, there has been incremental commercialization across a number of its services. For example, the outsourcing of Scottish Water's operations by way of private finance initiatives (PFIs) in the late 1990s has seen the private design, construction, operation and financing of 21 Wastewater Treatment Works (Kane and Mitchell, 2010). In early 2011 after years of debate over the merits of commercialization, the poor conditions of plants and inadequate quality of services were uncovered by University of Strathclyde researcher Tommy Kane by way of hundreds of Freedom of Information Inquiries (Edwards, 2011). It was revealed that not only have Scottish Water paid out over more than £100 million in assistance to private corporations to improve these operations, but also that Scottish Water will monitor the contracts and consider buying back these operations.

The prevailing relations of capitalist economies in many cases demand and in other cases compel people to enter into contractual relations with corporations, as is the case with commercialized water systems in Scotland and elsewhere. To separate "civil society" from the corporate sector on the basis of a normative distinction is to presumably separate ordinary people from the very conditions of their material existence. The interaction between people and corporations often determines the health (sustenance) and financial (wage-labour) wherewithal of ordinary individuals. Under a different or alternative mode of production this interaction may be different or may not even exist; however, in a society where the prevailing force behind the production of goods and services is the corporation (and therefore an elemental component of the totality of capitalist relations) it is disingenuous to separate people from this interaction, as uneven and unjust as this relation may be. Furthermore, this separation obscures the unequal property relations that facilitate the distribution of goods and services.

The argument here is that we should analyse society as it were and not as we may think it ought to be. As Marx argued, the "anatomy of civil society is to be sought in political economy" (Marx, 1976, p4). The political economy of water, as argued here, explores the underlying order that governs the global economy paying critical attention to the political and economic institutions and processes within which various actors, conditioned by the sphere of production and exchange and mediated by the prevailing property relations therein, interact with each other and the non-human world to construct our material existence, including access to water. In this way, contested relations comprise the very essence of civil society. The struggle over the right to water takes place in the terrain of civil society where transnational corporate private property increasingly precludes all other forms of rights, including the collective right to water. This is evidenced by the increasing rate of private sector involvement in water and wastewater services throughout the world. The materialist conception of civil society includes, then, as Castiglione states, "the totality of material relationships", or as Marx himself puts it, "the whole material intercourse of individuals within a definite state of the development of productive forces . . ." (Castiglione, 1994, p89; Marx, 1976, p163). This would include, necessarily then, individual activists, activist movements, non-profit organizations, non-governmental organizations, religious, professional and recreational organizations, but also lobbying groups (corporate and non-corporate) and corporations – increasingly in transnational form but also in their subordinated national character. To situate the struggle over water in a terrain of contested relations is to acknowledge the imbalance of power reflected in the unequal distribution of rights to water. As such it is to recognize that the struggle for collective rights is subordinated to the preeminent right of capitalist relations.

Conclusion

The political economy of water contextualizes the question of access as more a struggle over rights than over the biophysical matter of water itself. It is through an analysis of the unequal property relations that define the capitalist relations of production that we come to understand the complexities of the terrain of civil society and how, by recasting civil society as a sphere of contested relations, the prevailing property relations shape the contours of struggle and ultimately frame the outcomes of this struggle for ordinary people around the world every day.

What is truly at stake here is the disproportionate influence and power that corporations have over economic, political and legal resources and processes. Indeed, one may argue that in order to develop a just society where all receive water and wastewater services in accordance with need and not financial wherewithal, it is necessary to gain ownership and control over these spheres mentioned; however, this does not address the inherent inconsistencies and contradictions that presuppose the unequal relations in the first place, those elemental to the capitalist relations of production. Private property is the sine qua non of the capitalist system and principal social relation facilitated by the liberal democratic framework. These relations contradict the essence of collective life as the corresponding rights framework, facilitated and maintained by the liberal democratic state, is structured in such a way so as that rights sway in favour of some at the expense of others.

The success of the "right to water" movement will depend largely on how rights are framed and, ultimately realized. To follow the status quo is to perpetuate a structure of individual competing rights grounded in unequal property relations – the basis of all capitalist relations. This is made evident in the realization of human rights where, as Teeple (2005) notes, it would be hard to argue that when it comes to rights, human rights prevail instead of might is right. This said, what is required is a radical transformation in the mental conceptions of how we perceive not only our relations to water (and thus the non-human world in general) but also each other. Radical transformation in the way we conceive of rights in a capitalist society would challenge the very basis upon which this economic system is reproduced. Radical formulation of rights would subordinate individual rights in favour of the common right to a healthy, dignified life, including the right to water.

Notes

1 In *The Enigma of Capital* (2010) David Harvey refers to the contestation of forces interacting within the capitalist relations of production and he groups these into what he refers to as seven "activity spheres", including: "technologies and organisational forms; social relations; institutional and administrative arrangements; production and labour processes; relations to nature; the reproduction of daily life and of the species; and 'mental conceptions of the world'" (Harvey, 2010, p123). All of these "activity spheres" evolve on their own account but also in interaction with one another but not always simultaneously – Harvey suggests they are "co-present" and "co-evolving". Yet, all of the spheres are embedded in the "institutional arrangements" (i.e. prevailing property relations) that define the capitalist relations of production as well as the "administrative structures" (all levels of government) that facilitate the relations within. This framework, as Harvey notes, is taken from a footnote in *Capital Vol.1*. Of these activity spheres, the last – mental conceptions – refers to "expectations" and "beliefs" about the world and these are affected by "knowledge structures" and "cultural norms" and, in turn, have a tremendous effect on all of the other activity spheres, including the expansion and accumulation of capital which itself is effected by many of the spheres mentioned.

2 In *The German Ideology* (1845–46) Marx and Engels detail the relational aspect of human production beginning with what they refer to as the "first historical act" that being "the production of the means to satisfy . . . needs, the production of material life itself" (Marx, 1978, pp156–157). The first aspect of the first historical act, then, is the development of the instruments of production. The second aspect of this act is that satisfaction, "leads to new needs; and this production of new

needs is the first historical act". From this first historical act humans, "who remake their own life, begin to make other men, to propagate their kind".

3 Teeple argues that even though historical systems of exchange presupposed private property and the development of money, these relations still required notions of value, thus, "the development of money, latent" (Teeple, 1984, p137).

4 In *Capital Vol.1* chapter one Marx argues that there is nothing mysterious about the use-value of a commodity inasmuch as "by its properties it satisfies human needs, or that it first takes on these properties as the product of human labour" (Marx, 1976, p163). The mysteriousness of a commodity, then, according to Marx, appears as a result of the "suprasensible or social" embedded within the commodity form, that is, the totality of estranged and alienated social relations involved in the production of what appears at first sight a simple use-value but when taken to market mysteriously evolves into an exchange value. The real value of commodities, human labour, is alienated and instead the value of a commodity mysteriously appears as its exchange value.

5 Even where there is market failures some economists claim that market failures are a consequence of inadequate state intervention. Many economists believe that the state, in their role of re-regulation as opposed to regulation, should enable the necessary political and legal conditions in order to guarantee the success of the private sector in various sectors (i.e. providing an enabling environment). In much of their literature the UN, the World Bank and the World Water Forum stress the importance of state and local authorities' role in providing enabling mechanisms (social, political, legal and economic) that would contribute to the success of private industry.

6 Civil society was synonymous with the state and political society until the eighteenth century. Many commentators use the French Revolution as a historical departure when it was commonly agreed that the emerging and dominant private property relations necessitated a divorce between civil society and the state.

References

Amoore, L. and Langley, P. (2004) "Ambiguities of global civil society", *Review of International Studies*, vol 30, no 1, pp89–110

Anderson, T.L. and McChesney, F.S. (2003) *Property Rights: Cooperation, Conflict and Law*, Princeton University Press, Princeton, NJ

Anderson, T.L. and Simmons, R.T. (1993) *The Political Economy of Customs and Culture*, Rowman and Littlefield Press, Lanham, MD

Anheier, H.K. (2005) "Introducing the Journal of Civil Society: An editorial statement", *Journal of Civil Society*, vol 1, no 1, pp1–3

Arnold, C.A. (2002) "The reconstitution of property: Property as a web of interests", *Harvard Environmental Law Review*, vol 26, no 2, pp281–364

Bartelson, J. (2006) "Making sense of global civil society", *European Journal of International Relations*, vol 12, pp371–395

Bellamy, R.P. (2001) "Liberalism: Impact on social science", *International Encyclopedia of the Social & Behavioural Science*, pp8797–8801

Brown, T.C. (2009) "The time of globalization: Rethinking primitive accumulation", *Rethinking Marxism*, vol 21, no 4, pp571–584

Bryan, B. (2000) "Property as ontology: On Aboriginal and English understanding of ownership", *Canadian Journal of Law and Jurisprudence*, vol 13, no 1, pp3–31

Caruthers, B. and Ariovich, L. (2004) "The sociology of property rights", *Annual Review of Sociology*, vol 30, pp23–46

Castiglione, D. (1994) "History and theories of civil society: Outline of a contested paradigm", *Australian Journal of Politics and History*, vol 40, no 1, pp83–103

Chandhoke, N. (2002) "The limits of global civil society", in H. Anheier, M. Kaldor, M. Glasius (eds) *Global Civil Society*, Oxford University Press, Oxford

Chandler, D. (2004) "Building global civil society 'from below'?", *Millennium – Journal of International Studies*, vol 33, pp313–340

Cohen, J. and Arato, A. (1992) *Civil Society and Political Theory*, MIT Press: Cambridge, MA

Edwards, R. (2011) "This is damning evidence of the dangers of handing control of public services to private firms", *Herald Scotland*, 27 March. http://www. heraldscotland.com/news/home-news/this-is-damning-evidence-of-the-dangers-of-handing-control-of-public-services-to-private-firms-1.1092874?localLinksEnabled= false accessed 27 March 2011

Frow, J. (1996) "Information as gift and commodity", *New Left Review*, 219, pp89–108

Gellner, E. (1998) *Conditions of Liberty: Civil Society and its Rivals*, Penguin, Harmondsworth

Harvey, D. (2010) *The Enigma of Capital and the Crises of Capitalism*, Profile Books, London

Harvey, D. (2003) *The New Imperialism*, Oxford University Press, Oxford

Heritier, A. (2002) *Commons Goods: Reinventing European and International Governance*, Rowman and Littlefield Publishers Inc., Lanham, MD

Hinkelammert, F.J. (1986) *The Ideological Weapons of Death: A Theological Critique of Capitalism*, Orbis Books, Maryknoll, NY

Kaldor, M. (2003) "The idea of global civil society", *International Affairs*, vol 79, no 3, pp583–595

Kaldor, M. (2000) "'Civilising' globalisation? The implications of the 'Battle in Seattle'", *Millennium – Journal of International Studies*, vol 29, pp105–114

Kane, T. and Mitchell, K. (2010). "A steady flow of venality", *Scottish Left Review*, issue 60 September/October, pp25–27

Keane, J. (2003) *Global Civil Society?*, Cambridge University Press, Cambridge

Keck, M.E. and Sikkink, K. (1998) *Activists beyond Borders: Advocacy Networks in International Politics*, Cornell University Press, Ithaca, NY

Kumar, K. (1993) "Civil society: An inquiry into the usefulness of a historical term", *The British Journal of Sociology*, vol 44, no 3, pp375–395

MacPherson, C.B. (1978) *Property: Mainstream and Critical Positions*, University of Toronto Press, Toronto

Marx, K. (1978) in R.C. Tucker (ed) *The Marx and Engels Reader 2nd Edition*, W.W. Norton & Company Inc., New York

Marx, K. (1976) *Capital Volume One*, Vintage Books, New York

Miller, D. and Dinan, W. (2008) *A Century of Spin: How Public Relations Became the Cutting Edge of Corporate Power*, Pluto Press, London

Reeve, A. (1986) *Property*, Macmillan Education Ltd., London

Teeple, G. (2005) *The Riddle of Human Rights*, Garamond Press Ltd, Aurora

Teeple, G. (2000) *Globalization and the Decline of Social Reform: Into the Twenty-First Century*, Garamond Press Ltd, Aurora

Teeple, G. (1984) *Marx's Critique of Politics 1842–1847*, University of Toronto Press, Toronto

Walzer, M. (1995) "Introduction", in M. Walzer (ed) *Toward a Global Civil Society*, Berghahn Books, New York

White, G. (1994) "Civil Society, democratization and development: Clearing the analytical ground", *Democratization*, vol 1, no 3, pp375–390

6

SCARCE OR INSECURE?

The right to water and the ethics of global water governance

Jeremy J. Schmidt

Introduction

In 1977 a distinct shift occurred in global water governance. That year, at the UN Conference on Water in Mar del Plata, a stance was taken against the idea that water is abundant; a notion deemed too unscientific, too irrational and too normatively imbued to provide a grounding pro-position for managing water in industrial society (Biswas, 1978). In its place, and over the decades pursuant to the judgment of abundance, water scarcity and water security have emerged as the dominant propositions ordering the tasks of global water governance. As shown herein, this has substantively altered the context of governance by shifting it away from developing water for 'industrial society' and towards ordering water within the global economy. Further, these propositions contextualize, if they do not fully characterize, contemporary governance norms. As such, this chapter elucidates what the right to water must aim to accomplish if it is to present a viable and adequate response to the contemporary global water govern-ance milieu.

The first section argues that the categories of 'scarcity' and 'security' do not mirror the state of affairs currently besetting the Earth's socio-hydrological systems. Rather, they are judgments shared by a community of experts, academics and practitioners. These judgments legitimate claims to water by ordering the Earth's complex systems according to propositions amenable to, and in many cases constitutive of, a particular view of the world. Understood in this way, scarcity and security condition arguments about water allocation, use and the types of rights appropriate to protecting legitimate claims. Further, these propositions support norms (or rules of right conduct) that directly affect governance and which, because of water's central role in virtually all life projects, carry ethical consequences. The first section therefore positions scarcity and security in reference to both existing norms and the judgments that have legitimated the deployment of these propositions for the tasks of global water governance.

The second section provides a conceptual overview of how water scarcity and water security have been marshaled to order relationships between society and water. It considers how judgments regarding water's 'natural' state have become the operational norm for coordinating agreements in a manner that circumscribes governance arrangements across all communities. This has meant that water scarcity and water security ground the legitimacy of claims and the types of rights appropriate for protecting them. As such, both propositions have implications for the right to water. These implications, however, are far from uniform. As is argued below, this is due to these propositions being embedded in utilitarian accounts of community. The chapter concludes by arguing that we need to re-orient ourselves to an alternate account of community that can help us stay the problems of utilitarian governance. It shows why this is so by confronting the deficiencies of arguments regarding the 'water commons' and showing how alternate responses, such as deliberative democracy, are insufficient.

Governance propositions

Propositions have several functions, which include being expressive declarations (i.e. 'It's raining') that may be true or false. Further, they are understood to be that which is expressed regardless of how we say it (i.e. '*Il pleut*'). From certain philosophical perspectives, meaningful propositions are statements that reflect a particular state of affairs – meaning there are facts that verify, falsify or otherwise attest to their truth or falsity. For at least the first several decades of the 20th century such a view characterized much of scientific thought. Today this view is less palatable to many who see both science and policy as entangled in shared language games that structure the rules of truth, falsity and meaning (cf. Wittgenstein, 2001). As such, the truth-value of propositions is not simply based on whether they do or do not correspond to states of affairs. Hence, rather than look to what scarcity or security 'means' in an analytic sense (i.e. as reflecting states of affairs regarding the way things are) we may start by looking at the judgments that inform their acceptance and the norms that underlie them. Following Hannah Arendt (1982), judgments may be understood as relying on a public 'common sense' (*sensus communis*) which is developed through practice with others. Judgment is not just the teaching of how to reason, such as the correct logical steps needed to reach conclusions from premises. Rather, it is the ability to communicate with others in a way that anchors common meanings. Similarly, ethical judgments do not simply reflect states of affairs. And, further, they may never converge towards agreement based on 'the way things are' because they construe complex contexts in terms of how we ought to live, and not just in terms of what is (Williams, 1985). This section considers the communal, ethical judgments at work in global water governance.

Communal judgments

There is a good deal of interest in renewing the notion of community, especially as a means to rectifying the contemporary water governance milieu. While some employ the community concept to promote solidarity across cultural and national lines (i.e. Petrella, 2001), others are less optimistic. They see the global water governance community as comprised of an elite network that is largely occupied with pricing or otherwise privatizing water services (i.e. Goldman, 2009). In recent work, Bakker (2010) has moderated earlier criticisms of community as too 'ambiguous' to effectively ground water governance or rights in a way that resists privatization (i.e. Bakker, 2008). Thus, for instance, Bakker is hesitant about blanket support for 'community' just in case ambiguity leads to poor governance outcomes (see also chapter by Clark). Yet, for Bakker (2010), ambiguity can also be constructive if it works to disrupt normalizing or hegemonic narratives of 'modernist' governance programs. But 'modernist' states may be construed as communities too; and it may very well turn out that communities are relevant *regardless* of what scale of governance is considered (see generally Mason, 2000).

Communal arguments regarding the right to water start from the premise that existence is social. This strategy reflects the fact that, because water is constitutional for life, our relationship to it bears certain affinities with our relationship to each other (Illich, 1986; Strang, 2004; Shaw and Francis, 2008). Consequently, broadly held values are employed to assess the legitimacy of claims considered candidates for protection through rights. Put another way, we have a general water *ethos* of shared attitudes and norms upon which our water *ethic* is predicated, the latter of which we employ to test the validity of competing claims to water (cf. Chamberlain, 2008). Thus, instead of looking for analytical clarity, a communal perspective begins with the traditions of water governance practiced through different forms of social organization.

There are numerous examples wherein communal norms provide the basis for assessing particular claims to water: The *qanat* system of water distribution and governance in Iran emerged within the Zorastrian tradition that is now embedded in a largely Islamic perspective (Foltz, 2002; Balali et al, 2009). The French civil law tradition in the Canadian province of Quebec views water as a *res communis* and derives from Roman law and the community notion that water can neither be owned, nor sold (Cumyn, 2007). Some economists, such as Hanemann (2006), argue that riparian rights are essentially collective so as to capture the externalities of return flows in co-relative rights shared with others on the same stream. Even the doctrine of prior appropriation (first-in-time, first-in-right) has at its root a communal basis regarding fair distribution that was designed to prevent capitalist monopolies (Schorr, 2005). Interestingly, a shared feature of these and Hispanic governance traditions is that claims to water are often legitimated through mechanisms that prioritize community over efficiency; a recognition that '. . . water in place is a type of wealth' (Sax, 2010, p121). In this respect, all rights to water derive legitimacy from some level of community. Similarly, the right to water depends on a notion of community that is robust enough to order many competing water use traditions – from 'customary' to 'modernist' forms.

Legal scholars have also sought to reinvigorate the communal foundation of rights by showing how the stipulation that claims to water are relevant only between private individuals ignores empirical facts regarding both hydrological science and social institutions (Butler, 1986, 2000). These arguments emphasize that rights to water depend on communal acceptance for legitimacy and are therefore open to revision or rejection as social values change (Freyfogle, 1996). In this sense, the communal foundation for rights takes the idea that existence is social and connects it to the idea that rights to water are only legitimate when accepted by a group larger than that of the claimants (cf. Pradhan and Meinzen-Dick, 2003). Yet, at the level of *global* water governance, there is no larger group of claimants that exists to justify particular rights to water. Increasingly, this difficulty is handled through a model for water governance that is conceived of in polycentric terms (Ostrom, 2010). That is, power is understood as being held both within and across multiple scales of national and international actors. Thought of in this way, global governance norms are just those rules employed to coordinate and integrate various types and scales of communities.

The prevailing norms of contemporary global water governance derive from the western philosophy of utilitarianism (Feldman, 2007; Whiteley et al, 2008) and a western version of rationality (Wolf, 2008). Classic utilitarianism defined persons according to the liberal, individualist tradition of Mill (2001) and emerged as a means to finding an empirical measure of the good to achieve 'the greatest good for the greatest number'. In application, utilitarian governance has worked in tandem with the classic model of rationality based on universality and necessity, both of which are employed to eliminate judgments from reasoning (Brown, 1988). In regards to the former, classical rationality argues that individuals with the same information who reason correctly will reach the same conclusion. Thus, good reasoning holds to a standard independent of any particular person. In regards to the latter, a necessary connection between premises and conclusions '. . . allows us to understand *why* all rational individuals who start at the same point must arrive at the same conclusion' (Brown, 1988, p15). It is advantageous to employ this account in support of utilitarianism. For one, this version of rationality underlies much of modern science. For another, it enables utilitarianism to be contextually specific without being irrational (or relativistic) by a focus on consequences. So certain water uses could be good in one case, but bad in another, depending on their outcomes for enhancing the greater good.

Western techniques of utilitarian governance operate on a (claimed) synonymy between individual rationality and state resource policy by aggregating individual utility and defining 'community' as the sum. And, as utilitarian governance was gradually adopted as the *de facto* model for international water relations, it was expected that states would act in a manner congruent with accounts of individual rationality (see Blatter and Ingram, 2001). However, in practice the model of rationality breaks down (it also breaks down in epistemology [see Samuels et al, 2004] and in applied ethics [see Hoffmaster and Hooker, 2009]). For instance, although the model of agreement in international water treaties is claimed to be rational, actual agreements are often situated within deep-rooted value systems that are neither reducible to, nor commensurable with, western versions of rationality (cf. Wolf, 2008). This produces a dilemma. Either many international agreements are irrational or the standard of rationality is inadequate. If the latter, which eschewing a Eurocentric bias suggests, the neglected role of judgment may be reinvigorated. In this context, the 1977 judgment that 'water is scarce' was pivotal. This is the case because even though 'scarcity' was touted (*prima facie*) as objective, in operation it was a judgment that linked industrial societies to a de-politicized account of water's 'natural' distribution. Moreover, the global governance trend towards integrated water resources management (IWRM) is based on the proposition that 'water is scarce'. However, given the conceptual foundations of 'resource scarcity' in the utilitarian tradition of the west (Scott, 1998) and its habit of legitimating large interventions into the water cycle based on benefit-cost analysis (Feldman, 1991), there is a case to be made that 'integration' really means integrating water into the global political economy that supports capitalist societies and within which using economic instruments to allocate scarce resources is the primary (if not sole) medium for coordination.

Although the political economy of water governance is not the central topic here, it is a key feature in understanding *why*, of the possible suite of responses, the right to water has taken hold. One reason for this is that utilitarian forms of governance are reticent towards, and often outright reject, non-transferrable rights (depending on the version of utilitarianism), because respect for rights may confound attempts to achieve the greatest good. For instance, we may be compelled to respect the rights of a small group rather than using water in a way that would bring more benefits. So while in what follows I do not engage in the particular ways that forces of political economy 'dispossess' individuals of rights that do not fit capitalist rationality (see Swyngedouw, 2005) this is not because such contexts are unimportant. Quite the opposite; the focus here is to consider how utilitarian norms in political economy have been taken up in global water governance through the subtle, yet thorough-going repositioning of individuals from juridical to economic subjects across polycentric domains of power. This task is enabled through propositions that judge water to be 'scarce' or 'insecure' across all communities and, hence, across all subjects. The case of water is not unique in this respect (see Foucault, 2007). Yet, at the same time, countering this phenomenon requires new *kinds* of communal understandings in global water governance, a point returned to in the conclusion.

Judgments in global water governance

Philosophically, water governance is the process of linking propositions about what water is to policies regarding what it is for. In the practical rough-and-tumble of interagency coordination, partnerships and private sector participation, this process depends upon communal judgments which themselves rely on broadly held values and norms. For efforts that are global in scope, these judgments aim to coordinate action across a spectrum of divergent normative traditions. The complexity of such an extension necessitates a basis for coordination that is not shared by only a subset of the Earth's communities. Here the concept of an 'epistemic community' is helpful because global governance introduces new *kinds* of knowledge that must take root *somewhere* (cf. Hulme, 2010). An epistemic community '. . . is a network of professionals with recognized expertise and competence in a particular domain and an authoritative claim to policy-relevant knowledge within that domain or issue-area' (Haas, 1992, p3). As Haas (1992) notes, epistemic communities share: (1) Normative beliefs and values; (2) Causal beliefs regarding central problems in their domain of interest; (3) Perspectives of validity regarding knowledge; and, (4) Practices towards which they direct their efforts, such as those of human well-being. Despite the litigious aspects of building institutions for global water governance (see Conca, 2005), an epistemic community has emerged that fits with Haas' generic description. This section follows the conceptual developments of water scarcity and water security that proceed from it.

Water: scarcity and security

Since 1977, the world's water has been judged scarce. That year, the UN Conference on Water took aim at cavalier attitudes that considered water a 'free', discretionary resource. Thereafter, the central problems of water governance centered on ways to increase efficiency and provide ways of integrated decision making to correct the so-called 'inequitable distributions' of water (Biswas, 1978). Such an approach was necessary because the project of industrialized society required objective, scientific and technical expertise to manage water scarcity – not loose appeals to the claimed 'abundance' of water. Calculations of inequitable distributions were derived in relation to a Rational Planning model that relied on a comprehensive view of scientific and social facts to objectively optimize management outcomes (cf. Lindblom, 1999). *Pace* the classic model of rationality, the universal nature of water scarcity combined with a Rational Planning framework under the presumption that certain necessary outcomes of rectifying scarcity followed from the premises that contributed to it. In the words of the 'Policy Options' report from Mar del Plata, 'natural law dictates, man merely imitates' (Biswas, 1978, p72).

One of the problems of employing water scarcity as an input to Rational Planning exercises has been how to achieve the greatest good. As Delli Priscoli (1996, p30) has argued, international water management has been beset with a '... dialectic between two philosophical norms; one, the rational analytic model, often called the planning norm, and two, the utilitarian or free market model, often couched in terms of privatization.' It is in the reconciliation (or, alternately, the *reification*) of this dialectic that water scarcity moves from a 'natural' law to an ordering proposition for capitalist society. Thus, for instance, we can note that there is no convergence on how to define scarcity as reflecting certain states of affairs. In fact, the causes that link scarcity, water and humans may vary considerably depending on whether they are understood: in national terms of water-per-person, supply-demand scenarios, biophysical demands or adaptive capacity (see Rijsberman, 2006; Wolfe and Brooks, 2003). As noted earlier, when it comes to ethical judgments regarding how 'scarcity' fits with what we ought to do, it is unlikely that we will ever find convergence since *any* definition will lead to contestable outcomes. In this sense, the declaration in the Dublin Principles (1992) that water has an economic value in all of its competing uses has a special significance. This is because it links the proposition of scarcity to governance norms without that proposition reflecting a particular state of affairs. Rather, scarcity is the 'common sense' judgment of an expert community.

The Dublin Principles were commended (along with others) to world leaders at the UN Conference on Environment and Development in Rio de Janeiro (Rio). The main policy document issuing from Rio, *Agenda 21*, addressed concerns of water management in Chapter 18, which identifies fractured governance and the universal nature of water scarcity as rationale for IWRM. Even though Agenda 21 did not go very far in addressing scarcity, its international acceptance encouraged research programs that targeted the linkages among population demands, water scarcity metrics and their consequences for the aggregate economic, social and territorial ambitions of sovereign states (LeRoy, 1995). The neighboring concept of 'virtual water' likewise emerged in the early 1990s and linked scarcity, population demands (usually for food) and the global political economy (see Allan, 2003). In addition, the presumed tie between international policy coordination and the scarcity of water precipitated the development of policy networks, such as the World Water Council, that used international forums to coordinate and order the emerging global water governance community (see Giordano and Wolf, 2003).

An immediate upshot of the context in which water scarcity arose was tied to the rationality of employing scarcity to coordinate policy. In this regard, it was presumed that states were rational actors and that competition over scarce water could (and likely would) have direct implications on inter-state conflict and thereby on international security (Gleick, 1993). Hence, while the concept of water security is still nascent, one of the best ways to understand it is as a reflexive response to the problems issuing from the proposition that water is scarce. Following the utilitarian lead, the conceptual transition from scarcity to security augments universal claims regarding the inequitable distribution of water with concerns over local, place-specific issues by calculating disaggregated uses of water to ensure they are maximally beneficent, once summed, for the community.

In 1998, the World Water Council commissioned an independent report in preparation for the 2nd World Water Forum in The Hague, 2000. The report maintains continuity with the Dublin Principles, the conception of water as a scarce economic commodity and IWRM but it does so in the context of water security as part of its sustainable vision for the future (WWC, 2000). Security, the report suggests, comes through institutional renewal and the transfer of service provisions from the government to private enterprise. The new role for government is to adequately subsidize those unable to afford water and to act as a custodian of the environment. In this reorganization of water and service delivery, the most important policy recom-mendation the commission report offers is to adopt full-cost pricing to promote efficiency, accountability and conservation (WWC, 2000).

The recommendations of the World Water Council commission report were subsequently reflected in the Ministerial Declaration issued from The Hague (2000) entitled 'Water Security in the 21st Century'. It outlines the central vectors for the pursuit of water security: meeting basic needs, secur-ing the food supply, protecting ecosystems, sharing water resources, manag-ing risks, valuing water and governing wisely. Achieving targets in each area implies that water security should be thought of in terms of protecting and 'improving' natural assets, promoting sustainable development and provid-ing every person access to enough safe water at an affordable cost so that they can lead a healthy and productive life without threats from water-related hazards (i.e. floods) (Hague, 2000). Soon after, the International Conference on Freshwater convened in Berlin to make recommendations to the World Summit on Sustainable Development. The 2001 Berlin conference produced the 'Bonn Recommendations for Action' summarized as the 'Bonn Keys' of water security, decentralization, partnerships, cooperation, and governance. Reinforcing the water security agenda, the Bonn Recommendations (2001, p4) state that, 'Water should be equitably and sustainably allocated, firstly to basic human needs and then to the functioning of ecosystems and different economic uses including food security.' For its part, the United Nation's Development Program claims that we must move 'beyond scarcity' because the water crisis '... is holding back human progress, consigning large segments of humanity to lives of poverty, vulnerability and insecurity' (UNDP, 2006, p9). Critically, the idea that water security also has a shielding property has led to definitions that promote it as a concept for enhancing human productivity in a manner that protects vulnerable populations from certain types of polluted or ungoverned water, such as floods (see Grey and Sadoff, 2007).

The turn from scarcity to security reveals that 'inequitable distributions' of water are not gone, but they are no longer the driving problem. As water security evolves conceptually, its central metaphor is that it is the 'gossamer' that links food, energy, climate, economic growth and human security (World Economic Forum, 2009). As gossamer – finely stranded, flimsy and thin – water security is intended to keep the balance of human health and flourishing in tension with economic and environmental systems that can undergo rapid shifts (World Economic Forum, 2009). In a similar vein, Eckstein (2010) argues that, when water scarcity is coupled with uncertainty regarding climate variability, the task of addressing water security is primarily one of adaptation. Such adaptation should be based on a needs-based approach since formalizing a human right to water may outpace aquatic resources based on increased populations being legally entitled to more water (Eckstein, 2010). Finally, the Global Water Partnership (GWP, 2009) argues that water security must connect concern with economic activities to their effects on the resource base that underlie them. As these examples reveal, the conceptual turn to water security is primarily a reflexive response to the effects of human actions and not only the 'natural' state of water scarcity. Further, the conception of the state as synonymous (or very nearly so) with the aggregate of individual utility seekers within its jurisdictional purview links water security concerns across scales.

As the propositions of both water scarcity and water security evolve within the global epistemic community, it is critical to note that their uses change. This is not an analytically intractable problem for defining clear policy. Rather, it reflects the fact that these propositions are judgments that are used to order the Earth's complex socio-hydrological systems in a manner amenable to the project of coordinating governance in accordance with capitalist society. As these propositions ascend in policy discourse, they become constitutive parts of responding to the effects of ordering social relationships within the larger form of utilitarian rationality that is expected to obtain at multiple levels of governance – typically through a model of full-cost pricing that connects the conduct of individual utility seekers to those of nation states and multinational institutions promoting development through an aggregative model of community. In fact, as the first global, quantitative assessment of human water security and river biodiversity reveals, wealthy nations invest heavily against the effects of their own economic activities but tend not to re-order their activities so as to limit or confront their structural malignance (Vörösmarty et al, 2010). As the concluding portion of this chapter takes up, this reveals a key task for confronting the ethic of global water governance. What is needed is not a way to be more *reflexive*, but an ethic that enables *reflection* upon the judgments that currently order both what water is and what it is for.

Global governance and the right to water

In 2011, the World Economic Forum published *Water security: the water-food-energy-climate nexus*. That book resulted from an 'international, public-private-expert alliance' (World Economic Forum, 2011, pxvii) or what I've here termed an epistemic community. This work continues the effort to install water scarcity and water security as the primary propositions employed to circumscribe the range of governance arrangements considered legitimate for addressing global water problems. As argued above, within the context of utilitarian governance water has been viewed as a vital aspect to both the 'integration' of scarce resources and as the 'gossamer' that helps to mitigate the effects of capitalist societies. In such a context, a 'right to water' faces serious challenges since the basic categories ordering water are, in the main, those that emerge as responses to only one model of social organization. Given, and perhaps in spite of this milieu, there have been two major governance responses to the right to water: the 'water commons' and deliberative democracy. This conclusion shows some of the pitfalls of each with respect to their accounts of community and ends by arguing that we need to develop alternate *kinds* of communal norms to arrest the current categories ordering global water governance.

The water commons and deliberative democracy

In the literature on the commons, the constituency of the community is typically described vis-à-vis rational choice theory wherein individuals are self-interested, rational and strategically navigate social structures so as to procure resources in balance with long-term institutional viability (Ostrom, 1990, 2005). And, as Bakker (2010) and others in this book argue (see chapter by Mitchell), notions of the commons are therefore tied up with broader social, economic and political relationships. But as an entry point to community, it is important to recognize that the predominant accounts of the commons tend to see individuals as subject to the norms of rational-choice theory rather than as subject to the communal norms that guide the actual governance of water (Trawick, 2010). This helps explain why arguments for a human right to water that emerge from the 'water commons' that are criticized by Bakker (2007 and in this book) do not escape a certain congruence with the private rights that typify neo-liberal economic theory; both use an aggregative model that defines the community as the sum of individual util-ity seekers. More generally, the theoretical pitfalls of the 'commons' literature can be traced back to its conceptual roots in western ideas of property and an inability to adequately capture aspects of non-western governance rela-tionships because of the dualisms that it employs, such as seeing nature as inherently passive (see Schmidt and Dowsley, 2010). Nonetheless, the idea of the 'commons' is a powerful political construct, and increasingly so as communities respond to global water governance programs and resist con-formance to liberal versions of utilitarianism as the norm for participation (see Boelens et al, 2010).

As an alternate and supplementary form of engagement in the 'commons', the concept of deliberative democracy is often appealed to as a way to stay utilitarian governance norms. This concept also has a notion of community underlying it. In it, the community is populated by individuals that inter-subjectively share a large set of background assumptions – a lifeworld – that orders their particular form of life (Habermas, 1984a, b). The lifeworld pro-vides a basis for collective understanding while the rules that condition communication enable people who differ in their own personal convictions to engage in meaningful assessments of each other's claims. This 'commu-nicative rationality' buttresses accounts of reasoned consensus and has been referenced by those who seek a way to understand the public versus private spheres of environmental policy in general (Norton, 2005), and in water governance in specific (Bakker, 2010). Deliberative democracy also rejects the idea that all agents reason according to some abstract standard of ra-tionality. Rather, the rules of rationality are shaped by participation in a community and as such may change if context-altering events come about (Habermas, 1996). From this point of view, it is the shared submission to procedures amenable to building consensus (not calculating the 'greatest good') that lends legitimacy to water governance exercises. Yet there are difficulties with this view as well.

Deliberative, democratic legitimacy for a 'right to water' requires agree-ments on key terms that, in turn, depend on a large set of lifeworld factors – such as issues of access, duty, custom, legal tradition, symbolic values, etc. In this regard, the shared norms that organize 'lifeworlds' are much too complex to give an exhaustive account of water's place within them based on rules for communicative rationality. For instance, disagreements between western managers and indigenous peoples in the American southwest were not just over what was most 'rational' but due to incommensurable value orientations that affected what types of propositions, and which communal judgments, adequately ordered governance decisions (Espelund, 1998). In that case, as with others where entire ways of making up the world conflict, the basic judgments of different communities did not order the Earth's com-plex systems using categories that were reducible to, or accurately translatable within, competing worldviews. This implies that the 'community' built on inter-subjective rationality cannot always account for the ways in which competing groups appeal and contest the application of rules, nor for the fact that oppressed groups may learn the rules of their oppressors only to use them in a different way (Tully, 2008).

An alternate starting point

The right to water must take up a position that counters the prevailing norms that have led to the problems it seeks to address. The coordinating norms of global water governance have been primarily, if not exclusively utilitarian and these, in turn, have sought to install the propositions of water scarcity and water security as judgments of an epistemic community that uses these propositions to support governance norms across all scales of governance. Thus, one of the boldest points of entry for the right to water is to contest the idea that governance should be oriented around communities of aggregated individuals who (it is claimed) act according to abstract standards of rationality. The models of community that come out of the 'commons' and 'deliberative democracy' offer ways to confront each of these tenets in politically cogent ways, but do not, in my view, go far enough.

To reconceptualize the global water governance community we will need to turn away from community as the aggregated utility of discretely ordered individuals and towards a conception of both humans, nonhumans and biophysical systems as comprising an interdependent community. *That* community may have some of the characteristics identified in theories of the commons, deliberative democracy and economics – but it should not be constrained by them. Thus, we should not reject outright the role for economics as a *kind* of coordinating governance norm. Pricing water, for example, is entirely appropriate in many circumstances – but not as a uniform prescription for coordinating diverse and heterogeneous communities under the guise of utilitarian success. While I am sympathetic to the views of others in this book, I do not wish to endorse the notion that human-water relationships can be characterized reflexively such that the task of a right to water is to identify the 'kind' of relationships that currently obtain, such as the Lockean 'core' of modern political society (see chapter by Staddon et al), versus the kind that we should rethink such relationships through, such as Marx's 'species-being' (see chapter by Linton). There are many *kinds* of moral, political, familial, biophysical, ecological and social relationships that variably affect different communities. Thus, without emphasizing differences, it is not clear that there are any necessary or contingent reasons to constrain our potential responses to the negative outcomes of capitalist techniques for adjudicating water scarcity. Nor should we simply adopt water security as the new 'paradigm' for arresting its ills.

Different *kinds* of communities and communal judgments are available for testing claims to water. As such, we should reject the pretence that global governance propositions reflect 'states of affairs' regarding human-water relationships and focus arguments regarding the right to water on those norms that are judged acceptable for coordinating diverse communities. Once we acknowledge the public and ethical nature of these judgments – in a manner that respects how ethical judgments are unlikely to converge because of 'the ways things are' – it is equally clear that the predominance of capitalist relationships is not a sufficient basis for convergence regarding what the right to water needs to accomplish. In this sense, the return to the collective and communal basis of law and water doctrines represents a first step in recovering those norms that exist in many communities but which have been suppressed through the emphasis of one – typically utilitarian – kind of governance norm. The next is to de-center the entire community of human persons in a manner that respects the vital flows of water to which many forms of life have a communal claim, and which any particular right to water must respect.

Acknowledgements

The author thanks David Boyd, Frédéric Julien and the editors for comments. The Pierre Elliott Trudeau Foundation and SSHRC provided support. All defects are my own.

References

Allan, J.A.T. (2003) 'Virtual water – the water, food, and trade nexus: useful concept or misleading metaphor?' *Water International*, vol 28, pp4–10

Arendt, H. (1982) *Lectures on Kant's political philosophy* (R. Beiner, ed), University of Chicago Press, Chicago

Bakker, K. (2007) 'The "commons" versus the "commodity": alter-globalization, anti-privatization and the human right to water in the Global South,' *Antipode*, vol 39, pp431–455

Bakker, K. (2008) 'The ambiguity of community: debating alternatives to private-sector provision of urban water supply,' *Water Alternatives*, vol 1, pp236–252

Bakker, K. (2010) *Privatizing water: governance failure and the world's urban water crisis*, Cornell University Press, Ithaca

Balali, M.R., J. Keulartz & M. Korthals (2009) 'Reflexive water management in arid regions: the case of Iran,' *Environmental Values*, vol 18, pp91–112

Biswas, A.K. (ed) (1978) *United Nations water conference: summary and main documents*, Pergamon Press, Oxford

Blatter, J. & H. Ingram (eds) (2001) *Reflections on water: new approaches to transboundary conflict and cooperation*, MIT Press, Cambridge, MA

Boelens, R., D. Getches & A. Guevara-Gil (eds) (2010) *Out of the mainstream: water rights, politics and identity*, Earthscan, London

Bonn Recommendations (2001) 'Water – key to sustainable development: recommendations for action,' http://www.water-2001.de/outcome/bonn_recommendations.asp, accessed 29 November 2010

Brown, H. (1988) *Rationality*, Routledge, New York

Butler, L. (1986) 'Defining a water ethic through comprehensive reform: a suggested framework for analysis,' *University of Illinois Law Review*, vol 439, pp439–480

Butler, L. (2000) 'The pathology of property norms: living with nature's boundaries,' *Southern California Law Review*, vol 73, pp927–1016

Chamberlain, G. (2008) *Troubled waters: religion, ethics, and the global water crisis*, Rowman and Littlefield, Lanham, MD

Conca, K. (2005) *Governing water: contentious transnational politics and global institution building*, MIT Press, Cambridge, MA

Cumyn, M.C. (2007) 'The legal status of water in Quebec,' *Quebec Studies*, vol 42, pp7–15

Delli Priscoli, J. (1996) 'The development of transnational regimes for water resources management,' in Abu-zeid, M. & A. Biswas (eds), *River basin planning and management*, Oxford University Press, Calcutta, pp19–38

Dublin Statement (1992) 'The Dublin statement on water and sustainable development,' http://www.wmo.ch/pages/prog/hwrp/documents/english/icwedece.html, accessed 29 November 2010

Eckstein, G. (2010) 'Water scarcity, conflict, and security in a climate change world: challenges and opportunities for international law and policy,' *Wisconsin International Law Journal*, vol 27, pp410–461

Espelund, W.N. (1998) *The struggle for water: politics, rationality, and identity in the American Southwest*, University of Chicago Press, Chicago

Feldman, D. (1991) *Water resources management: in search of an environmental ethic*, Johns Hopkins University Press, Baltimore

Feldman, D. (2007) *Water policy for sustainable development*, Johns Hopkins University Press, Baltimore

Foltz, R.C. (2002) 'Iran's water crisis: cultural, political, and ethical dimensions,' *Journal of Agriculture and Environmental Ethics*, vol 15, pp357–380

Foucault, M. (2007) *Security, territory, population: lectures at the College de France 1977–1978*, Palgrave Macmillan, New York

Freyfogle, E. (1996) 'Water rights and the common wealth,' *Environmental Law*, vol 26, pp27–51

Giordano, M. & A.T. Wolf (2003) 'Sharing waters: post-Rio international water management,' *Natural Resources Forum*, vol 27, pp163–171

Gleick, P.H. (1993) 'Water and conflict: fresh water resources and international security,' *International Security*, vol 18, pp79–112

[GWP] Global Water Partnership (2009) 'Global water security: submission by the Global Water Partnership to ICE/RAE/CIWEM Report to Professor John Beddington, chief scientific advisor to HM Government,' 5

Goldman, M. (2009) 'Water for all! The phenomenal rise of transnational knowledge and policy networks,' in Kutting, G. & R.D. Lipschutz (eds) *Environmental governance: power and knowledge in a local-global world*, Routledge, New York, pp145–169

Grey, D. & C. Sadoff (2007) 'Sink or swim? Water security for growth and development,' *Water Policy*, vol 9, pp545–571

Haas, P.M. (1992) 'Introduction; epistemic communities and international policy coordination,' *International Organization*, vol 46, pp1–35

Habermas, J. (1984a) *Lifeworld and system: a critique of functionalist reason*, Beacon Press, Boston

Habermas, J. (1984b) *Reason and the rationalization of society*, Beacon Press, Boston

Habermas, J. (1996) *Between facts and norms: contributions to a discourse theory of law and democracy*, MIT Press, Cambridge, MA

Hague (2000) 'Ministerial declaration of The Hague on water security in the 21st century,' The Hague, the Netherlands

Hanemann, W.M. (2006) 'The economic conception of water,' in Rogers, P.P., M.R. Llamas & L. Martinez-Cortina (eds) *Water crisis: myth or reality?* Taylor and Francis, London, pp61–91

Hoffmaster, B. & C. Hooker (2009) 'How experience confronts ethics,' *Bioethics*, vol 23, pp214–225

Hulme, M. (2010) 'Problems with making and governing global kinds of knowledge,' *Global Environmental Change*, vol 20, pp558–564

Illich, I. (1986) *H₂O and the waters of forgetfulness*, Marion Boyars, London

LeRoy, P. (1995) 'Troubled waters: population and water scarcity,' *Colorado Journal of International Environmental Law*, vol 6, pp299–326

Lindblom, C. (1999) 'A century of planning,' in Meadowcroft, K.M. (ed) *Planning sustainability*, Routledge, New York, pp39–65

Mason, A. (2000) *Community, solidarity and belonging: levels of community and their normative significance*, Cambridge University Press, New York

Mill, J.S. (2001) *Utilitarianism*, Hackett Publishing Company, Indianapolis

Norton, B.G. (2005) *Sustainability: a philosophy for adaptive ecosystem management*, University of Chicago Press, Chicago

Ostrom, E. (1990) *Governing the commons: the evolution of institutions for collective action*, Cambridge University Press, New York

Ostrom, E. (2005) *Understanding institutional diversity*, Princeton University Press, Princeton

Ostrom, E. (2010) 'Polycentric systems for coping with collective action and global environmental change,' *Global Environmental Change*, vol 20, pp550–557

Petrella, R. (2001) *The water manifesto: arguments for a world water contract*, Fernwood Publishing, Halifax, Nova Scotia

Pradhan, R. & R. Meinzen-Dick (2003) 'Which rights are right? Water rights, culture, and underlying values,' *Water Nepal*, vol 9/10, pp37–61

Rijsberman, F.R. (2006) 'Water scarcity: fact or fiction?' *Agricultural Water Management*, vol 80, pp5–22

Samuels, R., S. Stich & L. Faucher (2004) 'Reason and rationality,' in Niiniluoto, I., M. Sintonen, & J. Wolenski (eds) *Handbook of epistemology*, Springer, the Netherlands, pp131–182

Sax, J.L. (2010) 'Understanding transfers: community rights and the privatization of water,' in Brown, P.G. & J.J. Schmidt (eds) *Water ethics: foundational readings for students and professionals*, Island Press, Washington, DC, pp117–124

Schmidt, J.J. & M. Dowsley (2010) 'Hunting with polar bears: problems with the passive properties of the commons,' *Human Ecology*, vol 38, pp377–387

Schorr, D. (2005) 'Appropriation as agrarianism: distributive justice in the creation of property rights,' *Ecology Law Quarterly*, vol 32, pp3–71

Scott, J. (1998) *Seeing like a state*, Yale University Press, New Haven

Shaw, S. & A. Francis (eds) (2008) *Deep blue: critical reflections on nature, religion and water*, Equinox, London

Strang, V. (2004) *The meaning of water*, Berg, New York

Swyngedouw, E. (2005) 'Dispossessing H$_2$O – the contested terrain of water privatisation,' *Capitalism, Nature, Socialism*, vol 16, pp1–18

Trawick, P. (2010) 'Encounters with the moral economy of water: general principles for successfully managing the commons,' in Brown, P.G. & J.J. Schmidt (eds) *Water ethics: foundational readings for students and professionals*, Island Press, Washington, DC, pp155–166

Tully, J. (2008) *Public philosophy in a new key* (2 vols), Cambridge University Press, Cambridge

[UNDP] United Nations Development Programme (2006) *Beyond scarcity: power, poverty and the global water crisis*, Palgrave Macmillan, New York, 50pp

Vörösmarty, C., et al (2010) 'Global threats to human water security and river biodiversity,' *Nature*, vol 467, pp555–561

Whiteley, J.M., H. Ingram & R. Perry (eds) (2008) *Water, place and equity*, MIT Press, Cambridge, MA

Williams, B. (1985, 2001) *Ethics and the limits of philosophy*, Harvard University Press, Cambridge, MA

Wittgenstein, L. (2001) *Philosophical investigations*, Blackwell, Malden, MA

Wolf, A. (2008) 'Healing the Enlightenment rift: rationality, spirituality and shared waters,' *Journal of International Affairs*, vol 61, pp51–73

Wolfe, S. & D. Brooks (2003) 'Water scarcity: an alternative view and its implications for polity and capacity building,' *Natural Resources Forum*, vol 27, pp99–107

World Economic Forum (2009) 'The bubble is close to bursting: a forecast of the main economic and geopolitical water issues likely to arise in the world during the next two decades.' Draft for discussion at the World Economic Forum Annual Meeting 2009

World Economic Forum (2011) *Water security: the water-food-energy-climate nexus*, Island Press, Washington, DC

[WWC] World Water Council (2000) 'A water secure world: vision for water, life and the environment,' *World water vision: commission report*, 70pp

7

THE RIGHT TO WATER AS THE RIGHT TO IDENTITY

Legal struggles of indigenous peoples of Aotearoa New Zealand

Jacinta Ruru

"Tuatahi ko te wai, tuarua whānau mai te tamaiti, ka puta ko te whenua"
When a child is born the water comes first, then the child, followed by the afterbirth

Introduction

According to the worldview of my ancestors, *wai* (water) is everything. To us, the Indigenous peoples of Aotearoa New Zealand,[1] known collectively as Māori, we live daily with this knowledge through our language, stories, songs, laws, and histories. As the *whakatauki* (proverb) above positions, water comes first. Or to give another example, in greeting someone new, we ask *"Ko wai koe?"* which queries "Who are you?" but more literally translates as "Who are your waters?" The answer will depend on which tribe and subtribe that person belongs to. For me, it is the Waikato River through my *whakapapa* (geneology) on my paternal grandfather's side (Ngati Raukawa). All tribes have these geographical identity markers linked to water (and also mountains). The link between land and water and humans is a common language feature. For instance, *iwi* means both tribe and bone; *hapu* means both subtribe and to be pregnant; *whanau* means both extended family and to give birth; *whenua* means land and afterbirth. *Wai* means water, but also memory and who.[2]

This chapter thus takes the opportunity to explore an Indigenous right to water not for drinking or sanitation purposes but for identity. This is because for the main part, access to clean water for survival is not an issue in Aotearoa New Zealand (although it is for other Indigenous peoples even in Western countries (e.g. see Morse, 2010)). Aotearoa New Zealand, a country that lies in the south-west Pacific ocean, has abundant freshwater (UNESCO, 2006) although issues of over-allocation of rights to take water for primarily agricultural and horticultural purposes and associated water pollution are mounting. The National-led Government is currently reviewing water law and policy, and is aware of the critical issue of how ought the law provide and recognise for Māori interests in freshwater. For instance, in a 2009-dated Cabinet paper it states:

> The rights and interests of Māori in New Zealand's freshwater resources remain undefined and unresolved, which is both a challenge and an opportunity in developing new water management and allocation models.

The Government-established Land and Water Forum recognises the importance of water to Māori. In its September 2010 report that outlines options for water reform, the Forum accepts that "Water is a taonga [treasure] which is central to all Māori life. Iwi interests in water are all encompassing, but key is their ability to maintain the health and wellbeing of waterways to sustain their tikanga – their way of life and being" (p7). Moreover, the Report acknowledges:

> Waterbodies frame iwi identity – tribal traditions are transmitted across generations by continuing customary practices with waterbodies and visions for the future of iwi turn on the health and wellbeing of freshwater. The obligation to protect freshwater and to maintain and express the spiritual and ancestral relationship with freshwater so as to leave a worthy inheritance for future generations is fundamental to iwi identity (p9).

While the Land and Water Forum Report does not attempt to define Māori tribal rights and interests in water, the report does acknowledge that for Māori "the contemporary discussion of freshwater evokes legacies of loss and exclusion and the denial of rights and responsibilities" (pvii).

This chapter thus focuses on exploring what rights in law Māori have to protect their relationship with water, and thus the health and wellbeing of both the water and themselves. Constitutional rights, general statutory rights, and specific negotiated rights form the structured basis of this chapter.[3] As this chapter concludes, while there has been legislative progress in recognising the Māori responsibility to water, the current Māori concerns are often trumped by other interests, such as the public need for hydro dams and sewage disposal, and the private need to take water for agricultural farming purposes.

No constitutional rights

Māori have no general constitutional rights to be recognised and heard within the court setting. This is in part because Aotearoa New Zealand does not have an entrenched constitution that recognises Māori rights (see Joseph, 2007). The Treaty of Waitangi is not part of the domestic law of Aotearoa New Zealand (see M. Palmer, 2008; G. Palmer, 1987). This is the document that the British Crown and Māori chiefs signed in 1840 that, according to the English version, Māori ceded sovereignty to the British Crown but Māori retained full exclusive and undisturbed possession of their lands, estates, forests, fisheries and other properties. Rather, it is commonly said that the Treaty forms part of our informal constitution along with the New Zealand Bill of Rights Act 1990 and the Constitution Act 1986. Therefore, for the judiciary or those acting under the law, the Treaty itself usually only becomes relevant if it has been expressly incorporated into statute. Even so, statutory incorporation of the Treaty of Waitangi has been a relatively recent phenomenon. Our legal history once endorsed the Treaty of Waitangi "as a simply nullity" (*Wi Parata v The Bishop of Wellington* (1877), 3 N.Z. Jur. (NS) 72, p78). Several statutes do now reference the Treaty, including the main statute that regulates rights to water as is discussed below. But before discussing domestic law, it is worthwhile mentioning possible recognition in international law.

In 2010, the New Zealand government announced that it supports, with conditions, the Declaration of the Rights of Indigenous Peoples. Article 25 specifically mentions water: "Indigenous peoples have the right to maintain and strengthen their distinctive spiritual relationship with their traditionally owned or otherwise occupied and used lands, territories, waters and coastal seas and other resources and to uphold their responsibilities to future generations in this regard". But this declaration is not binding on our domestic courts. The constitutional principle, to repeat, is one that positions Parliament as supreme. Māori thus have no general constitutional rights to water. Nonetheless, there is some domestic statutory recognition of the relationship Māori have with water as is now explored.

Some general statutory rights

The legislation

The central statute that manages the use of water in Aotearoa New Zealand is the Resource Management Act 1991 (RMA). The RMA is our pre-eminent natural resources statute (see Grinlinton, 2002; Nolan, 2005). It puts forward an all-encompassing regime for the sustainable management of land, air and water (see section 5). Central government retains some responsibility to influence this regime, primarily through setting national environmental standards, national policy standards and the New Zealand coastal policy statement. However, day to day control is vested in regional government and territorial authorities. These bodies prepare plans that contain rules concerning the use of land, air and water where appropriate, and stipulate when and where proposed activities may require resource consents to permit the use.

Thus the common starting point is that no person may do anything with land (including their privately owned land), air or water that contravenes a rule in a district plan unless the activity is expressly allowed by a resource consent, or coastal permit, granted by the territorial authority responsible for the plan, or a rule in a regional, or regional coastal, plan (sections 9–23; see Scott, 2006). The RMA gives regional and local councils the power to assert rules and guidelines for the take, use, damming, and diversion of freshwater (section 14). Regional councils have specific duties in regard to water. These include controlling the use of land for the purpose of the maintenance and enhancement of the quality and quantity of water in water bodies. The functions also include controlling the taking, use, damming and diversion of water for the purposes of setting maximum and minimum, and controlling the range of change of, water levels and flows. Regional councils need to control discharges of contaminants into water, and discharges of water into water. Regional councils can also, if appropriate, establish rules in a regional plan to allocate the taking or use of water, as long as the allocation does not affect the activities authorised in the RMA (section 14(3)(b)–(e)).

In formulating district and regional plan rules, and issuing resource consents, the RMA directs local authorities to recognise the Māori relationship with water. Section 6(e) mandates that all persons exercising functions and powers in relation to managing the use, development, and protection of natural and physical resources must recognise and provide for matters of national importance, including the relationship of Māori and their culture and traditions with water. However, this is one of several factors that local authorities must weigh in reaching decisions. Decision-makers must also recognise and provide for matters such as the protection of the natural character of the coastal environment, wetlands, lakes and rivers from inappropriate subdivision, use and development, and the protection of areas of significant indigenous vegetation and significant habitats of indigenous fauna.

Additionally, section 7(a) of the RMA directs that all persons exercising functions and powers in relation to managing the use, development, and protection of natural and physical resources, shall have particular regard to kaitiakitanga (the exercise of guardianship by Māori).[4] Again, it is one of several factors that must be considered. Section 7 lists several points ranging from the efficient use and development of natural and physical resources, the efficiency of the end use of energy, intrinsic values of ecosystems, and to the effects of climate change. Moreover, section 8 states that in achieving the purpose of the RMA, all persons exercising functions and powers under it, shall take into account the principles of the Treaty of Waitangi (Te Tiriti o Waitangi).

Sections 6(e), 7(a) and 8 provide a strong base for Māori to voice their concerns relating to the use of freshwater and to position the importance of water to identity. In addition, several other sections in the RMA create mandatory requirements on local authorities to listen to Māori. For example, in 2003, the RMA was amended to direct that a regional council, when preparing or changing a regional policy statement, must take into account any relevant planning document recognised by a tribal authority.[5] Section 62(1)(b) directs that a regional policy statement must state the resource management issues of significance to iwi authorities in the region. Moreover, since 2005, all local authorities must keep and maintain, for each iwi and hapu within its region or district, a record of their contact details and planning documents (section 35A(1)).

The RMA also provides for some substantial possibilities for Māori to be more actively involved in the governance of natural resources, including water. For example, the RMA empowers a local authority to transfer any one or more of its functions, powers, or duties to any iwi authority (section 32(2)). The RMA also enables a local authority to make a joint management agreement with an iwi authority and group that represents hapu for the purposes of the RMA.[6]

Hence, the picture put here is that there is a legal footing for Māori to have a right to be involved in the decision-making regarding the use and management of water and basis to position these rights in a cultural identity manner. How the courts have interpreted those rights is now discussed.

The court battles

In Aotearoa New Zealand, persons can appeal council decisions relating to issuing or not issuing resource consents to the Environment Court (K. Palmer, 2010). Thereafter, appeals are restricted to points of law, and go, in order, to the High Court, Court of Appeal, and lastly to the Supreme Court.[7] There are several instances where Māori, as objectors, have appealed council decisions that approved resource consents to take water, discharge wastewater into water, or dam water. In all of these cases Māori speak of the importance of the water to them culturally, including the belief that water has its own *mauri* (life force), and the importance of these places for food gathering, namely fishing.

In a survey of RMA cases concerning Māori and water, very few resulted in clear wins for Māori. One successful decision includes the recently decided *Heybridge* case where a developer sought resource consents for earthworks and water course modifications on its own land to create a four-lot subdivision (*Heybridge Developments Limited v Bay of Plenty Regional Council* [2010] NZEnvC 195). The local Māori opposed the application on the grounds that the site is of special significance including that it is the burial site of an important ancestor. The Environment Court agreed. It accepted (para 86):

that Pirirakau believe that serious cultural consequences would follow if the remains were disturbed during the development of the site. Those consequences include the potential loss of mana whenua [authority] over the land and waters that form this landscape should Tutereinga's remains be disturbed. This in customary terms would lead to the loss of identity of the hapu. They would, we were told, effectively cease to exist as a collective.

In other successful cases the courts also recognised the importance of that relationship between Māori and water. The Environment Court accepted Te Runanga o Ngai Tahu's argument that a man-made drain channel was subject to minimum flow requirements because the drain was linked to the Cust River and had capacity to support traditional use and values (*Federated Farmers of New Zealand (North Canterbury Province Inc) v Canterbury Regional Council* [2002] 8 ELRNZ 223). In another case, the Court agreed that the construction and operation of tidal gates on a tidal estuary would interfere with the natural flow of the tide causing its wairoa (spirit) to decay (*Ngataki, Ted and Ngati Tamaoho Trust v Auckland Regional Council* [2004] 9 NZED 725).

However, most cases do not result in clear wins for Māori. Several of the identified cases nonetheless did result in partial wins for Māori. A more recent one concerned the Bay of Plenty Regional Council reissuing to the Rotorua District Council a resource consent to take up to 3500 cubic metres of water per day over the summer months from the Taniwha Springs (*Te Maru o Ngati Rangiwewehi v Bay of Plenty Regional Council* [2008] ELRNZ 331). Ngati Rangiwewehi appealed the decision, relying on section 6(e) of the RMA, stating that their relationship with this water was a matter of national importance. The Court partially agreed by reducing the term of the resource consent from 25 to 10 years. The maximum daily volume and rate were not reduced.

Another case, decided back in 1996, involved the Mangakahia Māori Komiti challenging the resource consents issued to 17 dairy farmers to take water from the Opouteke River for irrigation (*Mangakahia Māori Komiti v Northland Regional Council* [1996] NZRMA 193). The Komiti contended that the water permits would adversely affect their right to catch fish in the river. The Court found a middle ground where the consents in most cases were slightly increased by one year, but the total level of water take permitted was reduced.

Kai Tahu were partially successful in a case where they argued that the issuing of resource consent to a particular jetboat company to operate ten additional trips on the Dart River for tourism purposes would adversely impact on their relationship with the river. The Court reduced the number of jetboat trips to four (See *G R Kemp and E A Billoud v Queenstown Lakes District Council* [2000] 7 NZRMA 289 and *Dart River Safaris Ltd v Kemp and Anor* [2000] NZRMA 440).

In another handful of cases, Māori were successful in protecting part of a lake from an aerial spray of weedkiller (*Walker v Hawke's Bay Regional Council* [2003] NZRMA 97), and a fish passage where consent had been granted for flashboards to be replaced by hydraulically controlled gates to manage a dam's water levels (*Mokau ki Runga Regional Management Committee v Waikato Regional Council* (Environment Court, Auckland, A046/06, 10 April 2006, Whiting J).

However, in some cases Māori lost their appeals outright. For example, Tautari lost their challenge of a resource consent that had been granted for the construction of a farm irrigation dam on the Waiopitotoi Stream. The consent also allowed the applicant to take up to 2700 cubic metres of water per day. Tautari appealed on behalf of the interests of the Māori people living only 6 km downstream of the proposed dam. Tautari argued that they had not been adequately consulted, the terms of the consent would disrupt the migration of traditional fish species and have a general effect on fish life as traditional sources of food. The Court disagreed (*Tautari v Northland Regional Council* [1996] 1 NZED 513).

In a case where Contact Energy appealed the Waikato Regional Council's refusal to grant resource consents for a proposed geothermal power station, Tauhara Middle Trust argued that the power station should not proceed because consultation had been inadequate and the geothermal resource is a *taonga* (treasure). The Environment Court found in favour of Contact Energy permitting the power station to be built (*Contact Energy Limited v Waikato Regional Council* [2000] 6 ELRNZ 1). In another case, Tainui argued that the resource consents issued to build a 86 ha engineered landfill would have potential adverse effects on the tributaries of the Waikato River, particularly the Clune Stream. The Court concluded that the existing conditions that the design is subject to "will adequately protect the Māori interests" (*Land Air Water Association v Waikato Regional Council* [2001] 7 NZED 26, para 494).

In another set of cases, Calter Holt Harvey was issued a 21-year term resource consent for discharge-to-water permits for its pulp and paper plant. Ngati Tuwharetoa did not agree with that term. The Environment Court had some sympathy for the tribe and held that they must be given more of a participatory role by reporting to them issues arising (*Te Runanga o Tuwharetoa ki Kawerau v Bay of Plenty Regional Council* [2002] 7 NZED 363). However, the High Court disagreed and held that Tuwharetoa have no consultation interests in the resource consent (*Calter Holt Harvey Ltd v Te Runanga o Tuwharetoa ki Kawerau* [2002] 8 NZED 335).

In the most recent case on water where a Māori tribe opposed the issuing of consents to Genesis Power to enable the Tongariro hydroelectric power development scheme to continue operating, the Māori tribe lost. Ngati Rangi Trust opposed the consents primarily because it involved diversion of water from the Whangaehu, Whanganui and Moawhango rivers in Lake Taupo and then into the Waikato River, and that their cultural traditions have been inhibited by a reduced flow of water, reduced water levels, degraded water quality and a change to the ecological system that affects the food chain in the water. While the Environment Court restricted the consents from 35 years to 10 years (*Ngati Rangi Trust v Manawatu-Wanganui Regional Council* (Environment Court, Auckland, A67/2004, 18 May 2004, Judge Whiting)), the High Court overruled that decision,[8] and the Court of Appeal has since endorsed the High Court's judgment.[9]

Thus, while the RMA does provide a platform for Māori to air their concerns, these concerns constitute just one of several factors that the decision-makers and the courts have to consider. The fact that Māori often lose in the courts is not because the courts lack the awareness of the importance of the RMA protections to Māori and their relationship with water. For example, Aotearoa New Zealand's then top appeal court, the Privy Council in the United Kingdom, stated in 2002, that sections 6(e), 7(a) and 8 provide "strong directions to be borne in mind at every stage of the planning process",[10] and that if alternative proposals exist that do not significantly affect Māori, then preference should be given to those alternatives even if they are not ideal. However, in that case, which concerned the laying of roads and not the take of water, Māori still lost. Moreover, the courts have been clear in stipulating that section 6(e) "does not create a right of veto"[11] for Māori and that it does not trump other matters.[12] In a more recent case, however, the Environment Court directed that section 6(e) "should not be given lip service to".[13] Nonetheless, overall the case law illustrates that while it is definitely a strong starting point to have legislative rights, those rights remain vulnerable and it requires significant time and resources on the part of Māori to pursue these rights. Another avenue has been to have recourse to the Treaty of Waitangi settlement process.

Some specific negotiated tribal rights

The Office of Treaty Settlements, situated within the Ministry of Justice, leads negotiations with specific tribes for redress (Office of Treaty Settlements, 2002). More than 15 legislated Crown-Māori claim settlement statutes have been now enacted, with many recognising the importance of freshwater to Māori identity, health and wellbeing. They all contain specific Crown apologies for wrongs done, financial and commercial redress, and redress recognising the claimant group's spiritual, cultural, historical or traditional associations with the natural environment. These statutes provide an additional avenue for Māori to advance their interests and connection to water than that offered through the RMA.

Many of the settlement statutes include provisions that acknowledge the importance of rivers and lakes to the tribes. For example, one of the early settlements, the Ngai Tahu Claims Settlement Act 1998, statutorily acknowledges the particular cultural, spiritual, historic, and traditional association of Ngai Tahu with the Mata-Au (Clutha) river (section 206). Schedule 40 of the Act records that the river is regarded as a descendant of the creation traditions where the earth mother and sky father separated to create a world of light in which we live in today. The schedule explains the importance of the river to Ngai Tahu ancestors including travels along it, camping beside it, and the collection of food from it. The schedule also speaks of the continuing mauri (life force) of the river, something that:

> represents the essence that binds the physical and spiritual elements
> of all things together, generating and upholding all life. All elements
> of the natural environment possess a life force, and all forms of life
> are related. Mauri is a critical element of the spiritual relationship
> of Ngai Tahu Whanui with the river.

Other settlement statutes include similar Crown acknowledgements of the importance of waterways to Māori. However, while statutory acknowledgements are important symbolically, they have little legal strength. If an area has a statutory acknowledgement attached to it, consent authorities must forward summaries of resource consent applications to the tribe. This at least gives the tribe knowledge of possible impending use of waterways, and the tribe can object to the proposed use but the consent authority does not have to give any particular weight to the fact that the statutory acknowledgement exists.

Some settlement statutes vest the fee simple estate of lakebeds in tribes. For example, in the Ngai Tahu Claims Settlement Act 1998, the bed of Te Waihora (Lake Ellesmere) was vested in Ngai Tahu (section 168). However, the Act expressly stipulates that this does not confer any rights of ownership, management or control of the waters or aquatic life in the lake to the tribe (section 171). The Act does provide for the possibility of a joint management plan for the lake that has since been developed and implemented (section 177).

The most recent, and by far the most revolutionary, settlements are those concerning the Waikato River – Aotearoa New Zealand's longest river (425 kilometres). The *Waikato-Tainui Raupatu Claims (Waikato River) Settlement Act 2010* has at its heart the Crown recognition that Waikato-Tainui regard the Waikato River as a tupuna (ancestor).[14] The Act endorses that a new vision and strategy "is intended by Parliament to be the primary direction-setting document for the Waikato River and activities within its catchment affecting the Waikato River" (section 5(1)). Key components of the Vision and strategy include: "(a) the restoration and protection of the health and wellbeing of the Waikato River; (b) the restoration and protection of the relationships of Waikato-Tainui with the Waikato River, including their economic, social, cultural, and spiritual relationships".[15]

The Waikato River Authority is the new statutory body responsible for setting the primary direction through the vision and strategy for the Waikato

River (section 22(2)). The Authority consists of ten members, including one member appointed from each of the iwi that link with the river (Te Arawa, Tuwharetoa, Raukawa, Maniapoto), a member appointed by the Waikato River Clean-up Trust, and five members appointed by the Minister for the Environment in consultation with other Ministers such as Finance, Local Government, Māori Affairs (schedule 6, clause 2(1)). The Act gives power to the new Waikato River Clean-up Trust. The Trust's primary object is "the restoration and protection of the health and wellbeing of the Waikato River for future generations" (section 32(3)). The Trust must also be involved in preparing a new integrated river management plan, along with relevant central departments, local authorities and other appropriate agencies (section 36(1)). The integrated river management plan must have conservation, fisheries, and regional council components (section 35(3)). Moreover, a joint management agreement must be in force between each local authority and the Trust in the near future (section 41(1); Te Aho, 2010). The *Waikato-Tainui Raupatu Claims (Waikato River) Settlement Act 2010* sets a significant standard of co-management between Māori and local authorities and encapsulates the importance of recognising and providing for the Māori relationship with water.

These tribes were successful in seeking this co-management settlement for several reasons. The settlement reflects the Crown's contemporary acceptance that at 1840, the date when the Treaty of Waitangi was signed, that specific Maori tribes possessed the Waikato River and that these tribes have consistently asserted their authority over the river. For example, the Act's preamble records an event in 1862 when a tribal chief warned the colonial Governor that the gunboat might not enter the river without his permission using these words: "E hara a Waikato awa i a te kuini, erangi no nga Maori anake". (The Waikato River does not belong to the Queen of England, it belongs only to Māori.) The Waikato River settlement was also partly an acknowledgement of the horrendous events that took place in the 1860s where the Crown's military forces declared war on the Waikato tribes. And then used the law to legitimate the confiscation of large tracts of land from the tribes including lands adjoining the Waikato River. As the settlement legislation accepts, the tribes were excluded from decision making concerning the river, and the Crown allowed the river to be used for:

> farming, coal mining, power generation schemes, the discharge of waste, and domestic and industrial abstraction. The wetlands were drained, flood protection schemes were initiated and sand and shingle were removed. While all of these uses of the Waikato River contributed to the economic growth of New Zealand, they also contributed to the pollution and deterioration of the health of the Waikato River and have significantly impacted on the fisheries and plant life of the River (preamble).

The 2010 settlement statute is reflective of the strong political will of these tribes to seek justice for these past wrongs. Other tribes are also seeking like justice and more legislated settlements are likely.[16]

Conclusion

The extent of Indigenous peoples' rights to govern, manage and even own freshwater is a topical issue in many countries. It is definitely hot in Aotearoa New Zealand. Since 1991, via the Resource Management Act 1991, decision-makers have had to recognise and provide for the relationship of Māori with their culture and traditions with water in regard to resource consent processes. While this has provided a right for Māori to be heard, their voice is often trumped by other public need-type interests, for example hydro dams and sewage disposal, or private agricultural demands for water. The new negotiated Treaty of Waitangi claim settlement statutes are reinforcing to decision-makers the importance of water to Māori and their health, wellbeing and identity. But again these recognition provisions are often not pitched in a mandatory form, although the recent Waikato River settlement indicates a movement towards giving Māori more managerial and governance responsibilities.

As the government continues to work through options for water reform, the unique challenge of confronting how best to create a legal governance regime that respects and provides for the Māori relationship to water is pressing. Hon. Pita Sharples, Minister of Māori Affairs, in 2009, positioned that governmental water reform "must be about the protection and preservation of water as a source of food, resources, and opportunities to maintain traditional connections and practices". With this being true, this chapter has thus focused not on exploring a right to water for drinking or sanitation purposes, but in laying out a new angle that centres on an Indigenous right to water *for identity*. The wellbeing of Māori depends on the wellbeing of water. With this sentiment, this chapter concludes with this well known Māori *whakatauki* (proverb):

"Ko te wai te ora ngā mea katoa"
Water is the life giver of all things

Notes

1 Aotearoa is a contemporary and commonly used Māori word for New Zealand. Māori is an official language of Aotearoa New Zealand.
2 For more information on the Māori language see H.W. Williams, 1971. For an introduction into the Māori worldview see Mead, 2003.
3 Note that some of the material for this chapter draws on work in Ruru, 2010.
4 Section 2 of the RMA defines kaitiakitanga to mean "the exercise of guardianship by the tangata whenua of an area in accordance with tikanga Māori in relation to natural and physical resources; and includes the ethic of stewardship".

5 Section 61(2A)(a) inserted by s 24(2) of the Resource Management Amendment Act 2003. Note that a similar direction exists for territorial authorities see: s 74(2A)(a) inserted by s 31(2) of the Resource Management Amendment Act 2003. Note that s 2 of the RMA defines an iwi authority as "the authority which represents an iwi and which is recognised by that iwi as having authority to do so".

6 Section 36B. See also ss 36C–36E. Note also that the Local Government Act 2002 similarly requires local authorities to have a certain level of regard to Māori and the Treaty of Waitangi.

7 Although note that prior to 2004, the Privy Council in London was New Zealand's last judicial bastion. See Supreme Court Act 2003.

8 *Genesis Power Ltd v Manawatu-Wanganui Regional Council* [2006] NZRMA 536. See also *Genesis Power Limited v Manawatu-Wanganui Regional Council* (High Court, Wellington, CIV-2004-485-1139, 22 May 2007, Wild J) and *Ngati Rangi Trust v Genesis Power Limited* [2007] NZCA 378.

9 *Ngati Rangi Trust v Genesis* (2009) 15 ELRNZ 164 (Court of Appeal, CA518/07, [2009] NZCA 222, 2 June 2009).

10 *McGuire v Hasting District Council* [2002] 2 NZLR 577, 594.

11 *Chief Executive of the Ministry of Agriculture and Forestry v Waikato Regional Council* 17 October 2006, Environment Court Auckland, A133/06, [49].

12 *Freda Pene Reweti Whanau Trust v Auckland Regional Council* 21 December 2004, Environment Court Auckland, A166/2004 – (2005) 11 ELRNZ 235, [50].

13 *Te Maru o Ngati Rangiwewehi v Bay of Plenty Regional Council* 25 August 2008 Environment Court Auckland A095/08 – (2008) 14 ELRNZ 331, [132]. For more discussion of this case see Bangma, 2009.

14 See Preamble (1) and 17(f), section 8(2) and (3). Note also the companion statute: Ngati Tuwharetoa, Raukawa, and Te Arawa River Iwi Waikato River Act 2010.

15 Schedule 2, clause 3. Note clause 3 is extensive and lists 13 objectives – only the first two are reproduced here.

16 For example, see the Nga Wai o Maniapoto (Waipa River) Bill, introduced into the House on 16 November 2010.

References

Bangma, W. (2009) "The decision in *Te Maru o Ngati Rangiwewehi* and the consideration of alternatives under the RMA", *Resource Management Journal*, p7.

Grinlinton, D. (2002) "Contemporary environmental law in New Zealand" in K. Bosselmann and D. Grinlinton (eds), *Environmental Law for a Sustainable Society*, New Zealand Centre for Environmental Law Monograph Series, vol 1, pp19–46

Joseph, P. (2007) *Constitutional Law and Administrative Law*, 3rd ed. Brookers, Wellington

Land and Water Forum (2010) *Report of the Land and Water Forum: A Fresh Start for Fresh Water*, Land and Water Forum, Wellington

Mead, H.M. (2003) *Tikanga Māori. Living by Māori Values*, Huia Publishers, Wellington

Morse, B.W. (2010) "Indigenous peoples and water rights: does the United Nations' adoption of the Declaration on the Rights of Indigenous Peoples help?", *Journal of Water Law*, vol 20, pp254–267

Nolan, N. (ed) (2005) *Environmental and Resource Management Law*, 3rd ed. LexisNexis, Wellington

Office of Treaty Settlements (2002) *Ka tika a muri, ka tika a mua. Healing the Past, Building a Future*, 2nd ed. Office of Treaty Settlements, Wellington

Palmer, G. (1987) *Unbridled Power: An Interpretation of New Zealand Constitution and Government*, 2nd ed. Oxford University Press, Auckland

Palmer, K. (2010) "Reflections on the history and role of the Environment Court in New Zealand", *Environmental Planning Law Journal*, vol 27, pp69–79

Palmer, M. (2008) *The Treaty of Waitangi in New Zealand's Law and Constitution*, Victoria University Press, Wellington

Ruru, J. (2010) "Undefined and unresolved: exploring Indigenous rights in Aotearoa New Zealand's freshwater legal regime", *Journal of Water Law*, vol 20, pp236–242

Scott, K.N. (2006) "From the lakes to the oceans: reforming water resource management regimes in New Zealand", *Journal of Water Law*, vol 17, pp231–245

Sharples, P. (2010) "Indigenous Peoples' Legal Water Forum" *press release*. Retrieved from http://www.beehive.govt.nz/speech/indigenous-peoples039-legal-water-forum

Te Aho, L. (2010) "Indigenous challenges to enhance freshwater governance and management in Aotearoa New Zealand – The Waikato River settlement", *Journal of Water Law*, vol 20, pp285–292

UNESCO (2006) The 2nd UN World Water Development Report. Retrieved from http://www.unesco.org/water/wwap/wwdr/

United Nations Declaration on Rights of Indigenous Peoples (2007) Retrieved from http://www.un.org/esa/socdev/unpfii/en/drip.html

Williams, H.W. (1971) *Dictionary of the Māori Language*, 7th ed. GP Publications Limited, Wellington

8

LEGAL PROTECTION OF THE RIGHT TO WATER IN THE EUROPEAN UNION

Marleen van Rijswick and Andrea Keessen

Introduction

In the European Union (EU), the coming into force of the Water Framework Directive (WFD) in 2000 marked a new approach in water management.[1] The Preamble to the Directive states that water is regarded as a heritage which must be protected, defended and treated as such. Water is explicitly considered not to be a commercial product like any other. Furthermore, the EU recognized the right to water in March 2010. This EU position does not necessarily reflect the position of the EU Member States. Not all have recognized the right to water. Some Member States regard water as a public or common good, while others have transformed water rights into property rights (Quesada, 2010; Dellapenna and Gupta, 2009; De Visser and Mbaziri, 2006; Kissling-Näf and Kuks, 2004). European citizens regard access to safe and clean water as important. The protection of water was one of the first topics to become regulated in the field of European environmental law (Jans and Vedder, 2008). Water quality in the EU has improved in the last few decades, but not enough. And increasingly, water scarcity is becoming an issue in the EU, especially in the south of Europe. In this regard it is important to realize that a lack of water and an eventual subsequent struggle for water does not so much take place between rich and poor or indigenous and 'new' inhabitants, but far more between different users of water, as water is used for drinking water, agriculture, energy, shipping, industry and recreation.

The European approach to the protection of the right to water is a combination of a human rights approach and integrated water resource management. This is logical in the sense that sufficient and clean water for all requires a sustainable and equitable use and the protection of water as a natural resource. Before we discuss the way the EU protects the right to water in more detail, some remarks on the system of EU law are given to ensure a proper understanding of the legal regime and the way public participation and access to justice – necessary elements to legally enforce the right to water – are arranged (see also the chapter by Staddon et al).

The EU as a supranational organization

The EU is neither a regular international organization nor a federal state. It is a supranational organization made up of 27 Member States and that has consequences for its organization and legal order. The EU legal order can be characterized as an integrated legal order (Jans et al, 2007). In the field of water law and policy, this means that the Member States, being France, Germany, Italy, Belgium, Luxembourg, the Netherlands (the six founding states of the European Community in 1951), Denmark, Ireland, the United Kingdom (accession in 1973), Greece (1980), Portugal, Spain (1986), Austria, Finland, Sweden (1995), Estonia, Latvia, Lithuania, Poland, the Czech Republic, Hungary, Slovakia, Slovenia, Cyprus, Malta (2005), Bulgaria (2007) share their responsibility with the EU. The European Union has the power to adopt binding legislation for all Member States and the power to enter into international agreements. When the EU is a party to an international agreement – for example, a Treaty or Convention in which the right to water is acknowledged – and it fulfils its international obligations by means of European legislation, international obligations become binding obligations for each Member State.

The EU regulates only what is absolutely necessary, which is based on the subsidiarity principle and the proportionality principle (Jans et al, 2007). In environmental policies, the most frequently used legal instrument is the directive, a piece of EU legislation that has to be implemented in the national legal orders of the different Member States in a way that guarantees that the objectives of the legislation are fully attained, while the choice of the means to realize them is to a large extent left to the Member States (Jans and Vedder, 2008). The Member States are accountable to the European Commission for compliance with their European obligations and can be brought before – and even sanctioned by – the European Court of Justice in case of non-compliance.

Because EU environmental directives should be transposed into national law by the Member States, the legal protection is based on the national law systems of the Member States. This follows from the principle of procedural autonomy. The boundary of this procedural autonomy lies in the obligation that legal protection has to be *effective*, which means that the national legal system has to guarantee that citizens may enjoy the full protection that EU law offers them. This means that when a European directive offers rights to citizens, these rights have to be implemented in binding legislation and it must be assured that citizens can enforce their rights before the national courts (Jans et al, 2007). This approach makes EU law more powerful than international law.

EU water law and the role of human rights

European water law has been based on a river basin management approach since the year 2000 (van Rijswick et al, 2010). It offers an integrated approach, with the aim of avoiding pollution, on the one hand, and promoting a sustainable and equitable use of water resources, on the other. These aims are influenced by the following guiding principles in EU environmental law. Substantive principles that are relevant for all environmental and water legislation are a high level of protection and improving the quality of the environment, sustainable development, the integration principle, the precautionary principle, the principle that preventive action should be taken, the principle that environmental damage should be rectified at the source, and the polluter (or user) should pay principle. For consumers this is important, because these principles protect the quality of drinking water resources and ensure that those who pollute or use most water will pay a proportionate part of the costs (Kaika, 2003).

EU water law is based on shared responsibilities between the EU and the individual Member States. EU water law contains goals that Member States should meet, but offers room for policy discretion for the Member States concerning the way these goals can be attained. The EU protects the right to water by a combination of human rights law and water law. When it comes to the protection of human rights and more specifically the right to water, international treaties and the European Convention on Human Rights play a more important role than EU environmental and water legislation.[2] However, since the reform of the European Union in 2010, human rights and environmental protection have been further encapsulated in the EU legal framework, being articulated in the Charter of Fundamental Rights, the EU Treaty and the Treaty on the functioning of the European Union.[3]

EU water law also has a procedural component, which ensures that citizens are informed and involved in planning and decision-making. These procedural rights can be found in the international Aarhus Convention and its implementation in European law.[4] Article 14 of the WFD contains more specific obligations with regard to public participation, which relate in particular to informing and consulting the public at large. According to the Court of Justice, these obligations must be implemented in the national law of the Member States.[5] These procedural rights are important because they give citizens the possibility to check if their interests are well protected, to further them during the decision-making stage and if necessary to enforce them before the courts.

Scope of the right to water

Worldwide, voices are calling for a 'right to water' to be seen as a human right (Smets, 2005; Filmer-Wilson, 2005). A right to water guarantees a given quantity of drinking water per individual, often combined with a right to sanitation. A common figure cited for levels at which this might be set is 50 litres per person per day, although there is considerable debate over whether this figure is sufficient or whether volumetric considerations are themselves problematic. Whatever, this is a very limited quantity of clean drinking water or water for domestic uses. The average European uses at least 175 litres a day. The right to water is increasingly being recognized in international conventions, but not to such an extent that a binding human right to water exists.[6] A broader scope of the right to water includes the protection of safety against flooding and sufficient, clean water for domestic use, sanitation (Smets, 2010), food production, energy supply and the protection of ecosystems (van Rijswick, 2008).

The right to (drinking) water can be inferred from various conventions which form part of the EU legal framework. Article 14(2) of the 1979 Convention on the Elimination of All Forms of Discrimination against Women compels states to ensure 'adequate living conditions, particularly in relation to . . . water supply' for women.[7] The 1989 Convention on the Rights of the Child compels states to combat disease and malnutrition 'through the provision of adequate nutritious food and clean drinking water' (Art. 24(2)).[8] The International Covenant on Economic, Social and Cultural Rights of the United Nations (ICESCR) is based on the Universal Declaration of Human Rights.[9] The ICESCR contains a basis for the right to water in Articles 11 and 12. Under Article 12(1) ICESCR everyone has the right to the enjoyment of the highest attainable standard of physical and mental health. Article 12(2) ICESCR stipulates that States Parties to the Covenant must improve all aspects of environmental and industrial hygiene and take steps to achieve the healthy development of the child.

According to the General Comment (no. 14) of the UN Committee of the ICESCR, Article 12 ICESCR refers not only to health care, but also to all other factors that determine the enjoyment of good health, such as access to clean drinking water, personal hygiene requirements, an adequate supply of safe food, and housing. Article 11(1) ICESCR – the right to an adequate standard of living – also covers the availability of drinking water. The definition of the right to water can be found in General Comment no. 15 on the Right to Water, adopted in 2002 by the Committee of Economic, Social and Cultural Rights. It states that the Human Right to water entitles everyone to sufficient, safe, acceptable, physically accessible and affordable water for personal and domestic uses. An adequate amount of safe water is necessary to prevent death from dehydration, to reduce the risk of water-related disease and to provide for consumption, cooking, personal and domestic hygienic requirements.

Implementing the internationally protected right to water therefore requires the following conditions to be met:

1. Availability: the supply of water for each individual must be adequate and continuous for personal and general uses, e.g. drinking, sanitation, washing clothes, preparing food and personal and household hygiene.
2. Quality: the water for personal and general use must be safe, and therefore free of micro-organisms, chemical substances and radiological hazards that are a danger to health. The colour, odour and taste of water must also be acceptable.
3. Accessibility: water and water facilities must be accessible to everyone, without discrimination. Accessibility comprises:

    - Physical accessibility: water and water facilities must be located within safe physical reach for all sections of the population. Sufficient, safe and acceptable water must be accessible within each household, school and workplace.
    - Economic accessibility: water must be affordable for everyone.
    - Equal accessibility: water must be accessible to all, including the most vulnerable and marginalized sections of society, with no conditions or penalties being attached.
    - Information accessibility: accessibility also covers the right to seek and receive independent information on water issues.

An important aspect of the accessibility of drinking water is economic accessibility. General Comment no. 15 does not require water to be free, but financial obstacles must not be such that they restrict accessibility. It states that any payment for water services has to be based on the principle of equity, ensuring that these services, whether privately or publicly provided, are affordable for all, including socially disadvantaged groups. Equity demands that poorer households should not be disproportionately burdened with water expenses as compared to richer households. Although a General Comment is not binding on the States Parties, this comment is regularly referred to.

For the actual protection of water rights it is necessary that individuals can rely on the right to water before a national court. This means within the EU that they depend on the inclusion of a provision from which a right to water can be derived in a treaty or agreement to which their state is a party or a national (constitutional) provision. They cannot rely on EU Directives, as none explicitly contains a right to water. They only elaborate aspects that are necessary to realize the right to water.

The WFD comes closest to incorporating the right to water as one of its objectives is to protect the quantity and quality of freshwater resources. Its Preamble asserts that good water quality will contribute to securing the drinking water supply. Article 1 WFD mentions as relevant goals a sustainable use of water and contributing to the provision of a sufficient supply of good quality surface water and groundwater as needed for sustainable, balanced and equitable water use. In addition, the Drinking Water Directive contains a responsibility for administrative authorities as well as for drinking water companies to ensure good quality drinking water.

Other aspects of a broader right to water can be derived from a range of EU Directives. The obligation to take care of proper sanitation derives from the Urban Waste Water Treatment Directive (Directive 91/271).[10] The protection of water for food production and economic activities derives from the WFD, while the protection of water necessary for the functioning of ecosystems is based on the WFD, the Birds Directive (Directive 79/409/EC) and the Habitat Directive (Directive 92/43/EC).[11] Finally, the protection against flooding is based on the Floods Directive (Directive 2007/60/EC).[12]

Recognition of the right to water by the EU

There is no separate human right to water embodied in European legislation (van Rijswick, 2011). The right to water is based on international commitments undertaken by the EU and is detailed in European water directives that have been enacted. In European law, fundamental rights and principles are closely interwoven. The right to (drinking) water can particularly be deduced from the general principles of EU law, the EU Charter of Fundamental Rights and the European Convention on Human Rights (ECHR).

On the basis of Article 6 of the Treaty on the Functioning of the EU (TFEU), the EU considers fundamental rights that have been granted on the basis of the ECHR to be general principles of Community law. Of relevance to water rights are both the substantive rights as laid down in Article 2 ECHR, which guarantees the protection of life, and Article 1 of the first Protocol to the ECHR (protection of property) as well as the procedural rights needed to realize the substantive rights. The latter can be found in Articles 6 and 13 ECHR (right of access to the courts).[13] With the entry into force of the Lisbon Treaty (amending the Treaty on European Union, and amending and renaming the EC Treaty), the European Convention on Human Rights and the Charter of Fundamental Rights of the European Union have gained importance. Also of importance for the substantive right to water is the London Protocol on Water and Health, as part of the Treaty of Helsinki,[14] and for the procedural rights the Aarhus Convention. The EU is a party to both treaties.

Even though the European Parliament and the Council of Europe support the right to water, a legally binding right has yet to be incorporated in a statutory text. On 22 March 2010 the EU took a tougher stance on the fundamental right to water:

> On World water day the EU reaffirms that all States bear human rights obligations regarding access to safe drinking water, which must be available, physically accessible, affordable and acceptable.

(. . .) The EU recognizes that the human rights obligations regarding access to safe drinking water and to sanitation are closely related with individual human rights – as the rights to housing, food and health. But even more than being related to individual rights, access to safe drinking water is a component element of the rights to an adequate standard of living and is closely related to human dignity. The principles of participation, non-discrimination and accountability are crucial. Water for personal and domestic use must be safe, therefore free from substances constituting a threat to a person's health. Access to adequate and safe sanitation constitutes one of the principal mechanisms for protecting the quality of drinking water.[15]

So the EU considers the fundamental right to water as an essential component of existing human rights and links its realization both to substantive and procedural obligations on the part of the Member States. However, even the restricted right to drinking water and sanitation is not (yet) recognized by all European Member States.[16] This means that not everywhere in Europe is it possible for citizens to ensure their right to water by commencing legal proceedings.

Case study of the right to water in the Netherlands

Since the effectuation of the right to water within the EU depends on the Member States, it is interesting to take a closer look at one of them. Like the EU, the Netherlands does not have an explicit right to water in its Constitution (*Grondwet, GW*) either. The right to water can be implied from Article 21 of the Constitution, the right to government care for keeping the country habitable and for protecting and improving the environment.[17] It can also be assumed that the right to water also constitutes part of the government's task to promote the health of the population (Art. 22 *GW*), and it may even be possible to link it to the inviolability of the physical person (Art. 11 *GW*). The Dutch right to water can thus be considered to be part of Dutch social fundamental rights, which should be seen as a duty of care on the part of the authorities and are intended as a task for the government to enact legislation.

In a case before the Maastricht district court, the right to water was for the first time explicitly recognized. The case was between a drinking water company and a citizen who had not paid his drinking water bill. Therefore the drinking water company refused to deliver any drinking water until the bill was paid. The court found that the defendant could not bypass the WML, the regional monopoly company for the supply of drinking water, to invoke his right to water. It found this right to be embodied in rights that have been codified and recognized by the Netherlands, especially the right to an adequate standard of living and the right to health (Articles 11 and 12 respectively of the International Covenant on Economic, Social and Cultural Rights). It also mentioned that the Netherlands had recognized the right to water at a session of the Human Rights Council in 2008. The defendant won the case, because the sum or arrears (around Euro 150) was too low to justify water being cut off.[18]

In a later ruling by the court of appeal in Den Bosch in a similar case, the court also recognized the existence of the right to water on the basis of a similar reasoning. However, in this case the defendant lost. Referring to General Comment no. 15, the appeal court held that the recognition of a human right to water does not mean that a claim can be made for the provision of water at no cost. It concluded that the right to suspend delivery is not in itself in conflict with the right to water.[19] A thousand litres of drinking water in the Netherlands costs around Euro 1.50, which can be deemed to be a reasonable price that does not exclude even vulnerable groups from the supply of drinking water.[20] Moreover, in addition to this relatively low price for drinking water, the available social security assistance to poor households in the Netherlands also ensures that citizens can realize their right to water.

Realization of the right to water in combination with an IWRM approach: protection of drinking water resources

Within the EU, the practical task of ensuring the right to water is embedded in an integrated water resources management (IWRM) approach (van Rijswick, 2011). The IWRM approach covers the protection of the resources of drinking water. Adequate protection of the quality of water resources reduces the necessity for further purification treatment of groundwater and surface water in order for the water to be used for consumption. IWRM is also about the sustainable use of water, including water to be used for the drinking water supply. Securing the supply of sufficient, safe and clean drinking water requires such an integrated approach.

The Drinking Water Directive (Directive 75/440/EEC), which established the protection of drinking water resources was one of Europe's first environmental Directives (it dates from 1975). It established quality requirements for the quality of fresh surface water which, after appropriate treatment, was to be used for the production of drinking water. This Directive has now been integrated into the WFD and was repealed as of 22 December 2007. However, the case law of the Court of Justice relating to the Drinking Water Directive 75/440/EEC is still of relevance for a correct interpretation and understanding of the WFD in so far as it protects the quality of drinking water.

The Drinking Water Directive set up a system of European and national quality standards by establishing limit values and target values for drinking water resources. Member States had to take all necessary measures to ensure that the water was in conformity with these values and the Directive was to be applied without distinction to national waters and waters crossing the frontiers of Member States. The approach of the Drinking Water Directive 75/440/EEC was programmatic. In order to achieve the objectives, Member States were to draw up systematic action plans, including a timetable for the improvement of the quality of surface water. If the quality of the surface water fell short of the mandatory limit values, it was, in principle, not to be used for the production of drinking water. The Member States were allowed to set stricter requirements; and the Directive also included a standstill principle. The various quality standards had to be transposed into binding national legal rules. In case of non-compliance with the Directive, third parties harmed by this non-compliance had to be able to rely on these mandatory rules before a court in order to be able to enforce their rights.[21]

The WFD established that the integration of the system for the protection of drinking water resources in the WFD should not lead to a lower protection regime. On the one hand, that is most certainly not the case, as the WFD even adds protective requirements. The WFD adds a quantitative element to the previous drinking water resource protection regime. It provides that the Member States have to protect, enhance and restore all bodies of groundwater, ensuring a balance between abstraction and the recharge of groundwater. An explicit quantitative requirement for surface water seems to be lacking, but this requirement is implied in the binding WFD obligation for surface water management to realize ecological objectives. That would be impossible without managing the abstraction of surface water. Moreover, the EU Drought Strategy (COM/ 2007/414) also encourages the sustainable use of water. The financial instruments as proposed by the WFD to ensure the cost recovery of water services are also expected to encourage the sustainable use of water. Indeed, the concern for a potential increase in the price of water among farmers made the provision on the payment of costs for water services a heavily debated one in the coming into being of the WFD and led to substantial weakening. Full cost recovery is not mandatory, but only something to be taken into account (Kaika, 2003).

On the other hand, the governance approach taken by the WFD (Scott and Holder, 2006; van Rijswick, 2008; van Rijswick et al, 2010) has resulted in a great deal of discussion on the legal status of the environmental objectives which are now mostly set by the Member States instead of by the EU (Keessen et al, 2010). This turmoil may have been unavoidable in so far as it concerns the new, ecological requirements (Howarth, 2006). Such requirements are arguably best set at lower levels than the EU level and perhaps cannot but be obligations of best effort as nature is unpredictable. However, the discussion on the legal status of the WFD objectives can easily contaminate the established obligation of result status of the old, chemical quality standards whose attainment is important for the quality of drinking water resources (Keessen et al, 2010). Indeed, the rules on hazardous substances and the legal character of the objectives of the WFD constituted major points of conflict in the drafting of the final text of the WFD (Kaika, 2003). Conclusive case law of the ECJ on the legal qualification of the objectives of the WFD is needed to put an end to this discussion.

Any undesirable developments regarding the protection of drinking water resources under the WFD might be offset by the newly introduced area-related provisions regarding drinking water resources. Under Article 6 WFD, all water bodies used for the abstraction of water intended for human consumption must be included in a national register of protected areas. These water bodies have to be explicitly identified and monitored (Article 7 WFD) and may be subject to a stricter protective regime. This will depend on how the WFD obligation is interpreted so that all Member States achieve compliance with any standards and objectives in 2015 at the latest unless otherwise specified in the European legislation under which the individual protected areas have been established. Arguably, this means that all objectives have to be met in good time in protective areas.

However, it is also possible that Member States may invoke exemptions in protected areas as well as elsewhere because the exemptions are part of the objectives. These exemptions are a delay in achieving the objectives, a lowering of the objectives, force majeure and changes or developments justified by overriding public interests. Or it may mean that Member States may invoke exemptions in protected areas in so far as the European legislation that established these areas offers exemptions. In the case of water bodies used for the abstraction of drinking water, the WFD provides that water must meet the requirements of the Drinking Water Directive 98/83 (see below). Article 9 of Directive 98/83 allows for a temporary derogation from the chemical quality standards, provided that it does not constitute a potential danger to human health and provided that the supply of water intended for human consumption in the area concerned cannot be maintained by any other reasonable means. If this interpretation of the WFD is correct, this derogation would then be the only justified reason for non-compliance with the WFD objectives for water bodies used for the abstraction of drinking water.[22]

Despite the formulation 'aim to achieve' good water status in the Preamble to the WFD, it is evident from the judgment of the Court of Justice in Case C-32/05 that the quality requirements relating to drinking water of Article 7 (2) WFD constitute obligations of result, because these obligations are formulated in a clear and unequivocal manner in order to ensure, in particular, that the water bodies of Member States meet the specific objectives laid down under Article 4 of the Directive. According to the Court, this provision thus imposes obligations as to the results to be achieved and must be transposed by means of measures having binding force. Member States must ensure the protection of the identified water bodies with the aim of avoiding any deterioration in their quality, in order to reduce the level of purification treatment required in the production of drinking water. Member States may establish safeguard zones for those bodies of water.

Protection of the quality of drinking water as a product

For the consumer, a realization of the right to water depends on the quality of the product and its supply (van Rijswick, 2011). The Drinking Water Directive (Directive 98/83/EC) regulates the quality of drinking water as a product and has remained in force despite the entry into force of the WFD.[23] It establishes a number of quality requirements for drinking water as a product, which can be supplemented by national law quality requirements. Only the duty to supply drinking water is not regulated by EU law. Provisions on the supply of drinking water in the WFD are limited to the requirement that Member States ensure an adequate contribution to the costs of water services, including the supply of drinking water (Howarth, 2009). Consequently, Member States are free to place the supply of drinking water in public or private hands, subject only to the EU and national regulations on the protection of the sources of drinking water, the quality of the product and the general competition rules.

Drinking water for human consumption obtained through the application of water treatment must meet the requirements of the Drinking Water Directive. The objective of this Directive is to protect human health from the adverse effects of any contamination of water intended for human consumption by ensuring that it is 'wholesome and clean'. The Member States must take the necessary measures to that effect. In accordance with the minimum requirements of the Directive, water intended for human consumption is wholesome and clean if it is free from any micro-organisms and parasites and any other substances in numbers or concentrations which constitute a potential danger to human health; if it meets the minimum requirements set out in Annex I, Parts A and B of the Directive; and if Member States take all other measures necessary to ensure that water intended for human consumption complies with the requirements of the Directive. The measures taken to implement the Directive may in no circumstances have the effect of allowing, directly or indirectly, either any deterioration of the present quality of water intended for human consumption so far as that is relevant for the protection of human health, or any increase in the pollution of waters used for the production of drinking water.

Here, too, quality requirements and corresponding monitoring must be established. Member States must adopt values which are applicable to water intended for human consumption for the parameters set out in Annex I. Annex I of the Directive lays down the limit values for these substances. Member States must set values for other additional parameters where this is necessary for the protection of human health within their territories or a part thereof.

Water supplied from a distribution network must comply with the parametric values as set out in the Directive, at the point, within premises or an establishment, at which the water emerges from the taps that are normally used for human consumption. In the case of water supplied from a tanker, it must comply with the parametric values at the point at which it emerges from the tanker, in the case of water put into bottles or containers intended for sale, at the point at which the water is put into the bottles or containers; and in the case of water used in a food production undertaking, at the point where the water is used in the undertaking.

Strict rules apply if the requirements are not met. In the case of water supplied from a distribution network, Member States are deemed to have fulfilled their obligations where it can be established that non-compliance with the quality requirements is due to the domestic distribution system or the maintenance thereof. This is different when it concerns premises and establishments where water is supplied to the public, such as schools, hospitals and restaurants. In such a situation and if there is a risk that water supplied from a distribution network might not comply with the quality requirements, Member States must nevertheless ensure that appropriate measures are taken to reduce or eliminate the risks, such as advising property owners of any possible remedial action they could take. Member States must ensure appropriate treatment techniques, installations and materials and have an obligation to inform and advise consumers. Informing the public is considered to be very important.

Comparing the international and European right to water

In this section we analyse the right to water as it is protected in EU law and assess whether it meets the constitutive elements of the right to (drinking) water in international law utilising the analytical framework developed by van Rijswick (van Rijswick, 2011). We can conclude that:

1   Availability is not formally guaranteed by an explicit provision in the Treaty mentioned with regard to the right to water, nor in the Charter of Fundamental Rights or the ECHR. It must be stated that, in practice, for almost all EU citizens water is available.
2   The requirements that water for personal and general use must be safe, and therefore free of micro-organisms, chemical substances and radiological hazards that are a danger to health, and that the colour, odour and taste of water must be acceptable is guaranteed by the Drinking Water Directive and enforceable before the national courts.
3   Accessibility, i.e. that water and water facilities must be accessible to everyone, without discrimination, is not formally guaranteed by European law. Citizens have to rely on the ECHR to enforce these rights. Nevertheless, in most EU Member States water is physically accessible, but

that may not be enough. Much will depend on the national legislation of the Member States. In a situation of drought and water scarcity the European Drought Strategy recommends that Member States place the supply of drinking water first in ranking. Member States can establish a water hierarchy for that purpose, and then drinking water should be placed at number one. Only when a national drought strategy is implemented in river basin management plans, the WFD provides for public participation requirements and legal enforcement before the national courts.

There is no formal legislation to protect economic accessibility, although the WFD prescribes that costs should be recovered from the users of a water service, which includes the supply of drinking water, and it can thus be deduced that more than a reasonable margin of profits should not be imposed on water service users. Equal accessibility can be deduced from the ECHR and the Charter of Fundamental Rights, because they provide that all European citizens should be treated equally. Information accessibility is taken care of in the European legislation that obliges the Member States to keep the public informed of environmental information, which implements the Aarhus Convention. This obligation is also reflected in the Water Framework Directive and the Drinking Water Directive.

Protecting the sources of drinking water through the Water Framework Directive imposes mandatory quality standards on the Member States, quantitative requirements relating to the management of groundwater resources and monitoring obligations. In addition, it encourages the use of other instruments, e.g. the creation of drinking water protection areas. Proportionate cost recovery for water services is not limited to water abstraction and water delivery, but also includes levies on water pollution as an instrument to reduce water pollution in addition to the existing regulation of water pollution. The European protection of the resources of the drinking water regime also benefits from the general public participation and judicial protection requirements that characterize European environmental law.

In addition to these aspects we would like to mention the legal regime of compensation for damages, which can be caused by the wrongful and lawful acts of the government. An example of the last category may be a reallocation of water use to protect the right to water of certain citizens which may harm existing users of the water. Finally, the Environmental Liability Directive (Directive 2004/35/EC) is relevant for the protection of water resources. This Directive provides for fault or negligence-based liability for damage to the environment. Its purpose is the prevention of damage and the provision of remedies once harm has occurred.

This brings us to the conclusion that the protection and enforcement of the core right to water does not fully meet the requirements set out in international law. On the other hand, in some aspects the European right to water has a broader meaning than usual in the international context. One should also realize that legal remedies against distribution measures, permits, plans, the costs of drinking water supply etc. depend on the national courts of the Member States, which may lead to different protection levels within the EU.

Concluding remarks

The European approach to integrated water management protects the right to water to a great extent, and perhaps even better than a formal right which cannot be enforced. European law tries to facilitate a just distribution of water rights and duties, but not in a clear way. There is room for improvement when it comes to the formal recognition of the right to water and to the transparency of the supply of information and the involvement of the public in decision-making under the current legal framework. Although the EU recognizes the right to water, it can only be enforced by means of classical human rights as protected in the ECHR and the Charter of Fundamental Rights. The combined approach of protecting the right to water by a human rights-based approach and IWRM ensures that there is not only a right, on paper, to water but also a real possibility of enjoying the right to water through the availability of sufficient and clean water. Without sufficient clean water, a right to water is an illusion.

In developed and industrialized states like the EU Member States the ambition to protect the right to water should be extended beyond the protection of a small amount of drinking water. For this purpose, first of all the quality and status of the norms that determine the scope of water rights should be clear. Even more important is clarity concerning the position and the balance struck between several aspects of (the broader) right to water. After all, that is what the European water struggle is all about. Thus, the allocation of water for domestic use and other than domestic uses should be fair, i.e. transparent and legitimate. A lack of transparency makes it difficult to judge whether the allocation of water has occurred in a fair way. The legitimacy of the allocation of water to various uses and the protection of the right to (drinking) water is served by taking public participation seriously and ensuring that legal remedies are available for everyone to enforce their rights.

Notes

1 Directive 2000/60 of the European Parliament and of the Council of 23 October 2000 establishing a framework for Community action in the field of water policy, OJ 2000 L 327/1. See also: http://ec.europa.eu/environment/index_en.htm.
2 Available at: http://www.echr.coe.int/echr.
3 Consolidated versions of the Charter of Fundamental Rights, the EU Treaty and the Treaty on the functioning of the European Union, OJ 2010 C 83. See also: http://eurlex.europa.eu/en/treaties/index.htm.
4 Aarhus Convention: Convention on Access to Information, Public Participation in Decision-Making and Access to Justice in Environmental Matters, available at: http://www.unece.org/env/pp/. The EU acceded to this Convention (Decision

2005/370/EC). Concerning access to information and public participation two Directives were adopted, Directive 2003/4/EC and Directive 2003/35/EC. Although both contain provisions on access to justice, a proposal for a Directive on access to justice has remained a proposal. A Regulation to apply the provisions of the Directive to EU institutions and bodies was adopted in 2006 (Regulation 1367/ 2006/EC). See: http://ec.europa.eu/environment/aarhus/index.htm#legislation.

5 ECJ, Case C-32/05 (Commission/Luxembourg) ECR 2006, I-11323.
6 For example, in the International Covenant on Economic, Social and Cultural Rights (1966) and as a fundamental right recognized by the General Assembly of the United Nations (A/RES/54/175) and defined in General Comment no. 15 on the Right to Water, adopted in 2002 by the Committee of Economic, Social and Cultural Rights.
7 See: http://www2.ohchr.org/english/bodies/cedaw/index.htm.
8 See: http://www2.ohchr.org/english/law/crc.htm.
9 Respectively available at: http://www2.ohchr.org/english/law/cescr.htm and http:// www.ohchr.org/EN/UDHR/Pages/Introduction.aspx.
10 Council Directive 91/271/EEC of 21 May 1991 on Urban Waste Water Treatment, OJ 1991 L135/40.
11 Directive 2009/147/EC of the European Parliament and of the Council of 30 November 2009 on the conservation of wild birds (codified version) OJ 2010 L 20/7 and Council Directive 92/43/EEC of 21 May 1992 on the conservation of natural habitats and wild fauna and flora OJ 1992 L 206/7.
12 Directive 2007/60 of the European Parliament and of the Council of 23 October 2007 on the assessment and management of flood risks, OJ 2007 L 288/27.
13 See, for example, ECHR 20 March 2008 (deaths as a result of mudslides in Russia, violation of Art. 2 ECHR); ECHR 27 January 2009 (a violation of Art. 8 ECHR because the government did not provide the people living in the neigh- bourhood with sufficient information about the risks associated with a company that used cyanide to extract gold); ECHR 12 November 2006 (a violation of Art. 8 ECHR); ECHR 30 November 2004 (Oneryildiz, a violation of Art. 2 ECHR); ECHR 10 November 2004 (Taskin, a violation of Art. 8 ECHR); ECHR 16 November 2004 (Moreno Gomez, a violation of Art. 8 ECHR); ECHR 9 June 2005 (Fadeyeva, a violation of Art. 8 ECHR).
14 Treaty of Helsinki, Convention on the Protection and Use of Transboundary Watercourses and International Lakes (UNECE Water Convention), available at: http://www.unece.org/env/water/text/text.htm.
15 Brussels, 22 March 2010, 7810/10 (Presse 72) P12/10.
16 EU Member States that have recognized a fundamental right to water are Belgium, France, Finland, Germany, Italy, Norway, Portugal, Spain, Sweden, Switzerland, Ukraine, the Netherlands and the United Kingdom. The right to water is also recognized in a number of Latin American and African countries.
17 Parliamentary Proceedings II 2005/06, 21 501–30, no. 137, p. 3.
18 District court of Maastricht, subdistrict section (Heerlen), 25 June 2008, no. 294698 CV EXPL 08-4233, LJN BD5759, NJCM-bulletin 2009, pp. 249–255.
19 Court of appeal of Den Bosch, 5 March 2010, LJN BL6583.
20 In a similar vein: District court of Groningen, 19 February 2010, LJN BL4579.
21 ECJ, Case C-60/01 ECR 2002 I-05679; ECJ, Case C-266/99 ECR 2001 I-01981; ECJ, Case C-56/90 ECR 1993 I-04109; ECJ, Case C-92/96 ECR 1998, I-00505; ECJ, Case C-337/89 ECR 1992 I-06103; ECJ, Case C-316/00 ECR 2002 I-10527; ECJ, Case C-147/07 OJ 2008 C 79/8; ECJ, Case C 58/89 ECR 1991 I-4983.
22 This discussion led to several articles in the Dutch environmental law journal *Milieu en Recht*.
23 Council Directive 98/83/EC of 3 November 1998 on the quality of water intended for human consumption, OJ 1998 L 330/32.

References

Dellapenna, J.W. and Gupta, J. (eds) (2009) *The Evolution of the Law and Politics of Water*, Springer.

Filmer-Wilson, E. (2005) The human rights-based approach to development: the right to water, *Netherlands Quarterly of Human Rights*, vol. 23, no. 2, pp. 213–241.

Howarth, W. (2006) The progression towards ecological quality standards, *Journal of Environmental Law*, vol. 18, no. 1, pp. 3–35.

Howarth, W. (2009) Cost recovery for water services and the polluter pays principles, *ERA Forum*, Springer Verlag, pp. 565–587.

Jans, J.H. and Vedder, H.B. (2008) *European Environmental Law*, Europa Law Publishing, Groningen.

Jans, J.H., De Lange, R., Prechal, S. and Widdershoven, R.J.G.M. (2007) *Europeanisation of Public Law*, Europa Law Publishing, Groningen.

Kaika, M. (2003) The Water Framework Directive: a new directive for a changing social, political and economic European framework, *European Planning Studies*, vol. 11, no. 3.

Keessen, A.M., Van Kempen, J.J.H., van Rijswick, H.F.M.W., Robbe, J. and Backes, C.W. (2010) European river basin districts: are they swimming in the same implementation pool?, *Journal of Environmental Law*, vol. 22, no. 2, pp. 197–222.

Kissling-Näf, I. and Kuks, S. (eds) (2004) *The Evolution of National Water Regimes in Europe, Transitions in Water Rights and Water Policies*, Kluwer Academic Publishers, Environment and Policy, vol. 40.

Quesada, M.N. (2010) *Water and Sanitation Services in Europe, Do Legal Frameworks Provide for 'Good Governance'?*, UNESCO Centre for Water Law, Policy and Science, University of Dundee.

van Rijswick, H.F.M.W., Gilissen, H.K. and Van Kempen, J.J.H. (2010) The need for international and regional transboundary cooperation in European river basin management as a result of a governance approach in water law, *ERA Forum*, Springer Verlag, pp. 129–157.

van Rijswick, H.F.M.W. (2008) *Moving Water and the Law: On the Distribution of Water Rights and Water Duties in European and Dutch Water Law*, Europa Law Publishing, Groningen.

van Rijswick, H.F.M.W. (2011) The status of consumers in European water regulation, *European Journal of Consumer Law / Revue Européenne de droit de la consummation*, 2011/1, pp. 115–148.

Scott, J. and Holder, J. (2006) Law and environmental governance in the European Union, in: G. de Burca and J. Scott (eds), *Law and New Governance in the EU and the US*, Hart Publishing, Portland.

Smets, H. (ed.) (2005) *Le droit à l' eau dans les legislations nationals / the right to water in national legislations*, Académie de l'eau, Paris.

Smets, H. (ed.) (2010) *L'accès à l'assainissement, un droit fondamental / the right to sanitation in national laws*, Editions Johanet, Paris.

De Visser, J.W. and Mbaziri, C. (eds) (2006) *Water Delivery: Public or Private?* Utrecht University, Centre for Environmental Law.

9

RIGHTS, CITIZENSHIP
AND TERRITORY

Water politics in the West Bank

Ilaria Giglioli

Introduction

In recent years, the concept of the right to water has gained currency both
as a basis of local mobilisation, and as a buzzword in international policy
fora. While this concept garners consensus at a broad theoretical level, it
often becomes contentious in debates around its concrete implementation.
In the Palestinian case, however, even very abstract notions of the right to
water are highly contested. While all parties involved in the local and international debate on water in the Occupied Palestinian Territories recognise
the severe lack of access to water for Palestinians, this consensus is broken
if the issue is addressed in terms of Palestinian right to water. The strongest
reluctance to use this language comes from the Israeli side, however many
international co-operation organisations present in the territories also refrain
from engaging in a rights discourse, preferring instead to focus on the tech-
nicalities of water infrastructure development and water sector management.
Where the rights discourse does emerge, most recently in the context of a
campaign push by various local and international human rights organisations, it has been in the framework of a human right to water. According
to this framework, Palestinian right claims to water stem from their natural
rights as human beings rather than from their political status as legitimate
members of a polity with its own rights of self-determination over natural
resources. The latter position – which is held by some Palestinian voices – is
particularly contentious as it is strongly linked to broader Palestinian territorial and citizenship claims.

   This chapter explores the specific reasons that render the right to water a highly
contested concept in contemporary Palestine.[1] The discussion of the Palestinian
case is grounded in a conceptual exploration of the relationship between
rights, citizenship and access to water in different historical-geographical

contexts. In the light of this, the chapter then turns to explore the historical unfolding of water resource development in the Palestinian West Bank in the course of the 20th century, focusing on the shift of actors and related philosophies of water resource development. As each one of the five different political powers who have been in control of the West Bank over the course of the 20th century captured and channelled the water resources of the area according to their immediate strategic necessities and their visions for the future of the region, these changes also brought about shifts in the type of claims which were made by various parties over access to water resources, and thus in the nature of conflicts around water. Particularly since the Israeli occupation of the West Bank in 1967, conflicts around access to water became essentially conflated as conflicts around self-determination over natural resources, a configuration which lies at the basis of the current highly contested notion of a Palestinian right to water.

Water, rights and citizenship

In the course of the 20th century, water network development was increasingly carried out by the state, which was thus able to extend its material presence over national territory and resources on one hand, and on the other to channel the distribution of these resources and regulate access to them. Throughout different historical/geographical contexts, access to water was thus strongly determined by the type of relationship that citizens established with the state, and discrimination between different levels of access to water was strongly linked to different levels of state recognition. In this section we will examine some of the different contexts that are useful to elucidate the changing relationship between the authorities responsible for the development of water resources and Palestinian residents of the West Bank.

In Western Europe and North America of the second half of the 20th century, the development of water infrastructure occurred in the context of large-scale centralised planning, and generalised access to water represented a fundamental 'emblem of inclusionary citizenship' (Bakker, 2005, p559), an essential economic and social right through which the state asserted its legitimacy towards its citizens (Bakker, 2003a). While the most marginal sectors of the population, such as indigenous communities in Canada (Simeone, 2009), were not always reached by networks, exclusion was not an explicit policy, and the general principle was one of access to water for all.

In other contexts, however, selective channelling of water between different sectors of the population corresponded to an explicit policy of differentiation. This was particularly clear in many cities developed under European colonialism, in which the colonial and native quarters of the city were distinguished by substantially different levels of water infrastructure provision. This simultaneously ensured an uneven level of access to water between different parts of the population, and symbolically marked a difference of civilisation between a European, clean and modern population and an indigenous, primitive and unhealthy one (Gandy, 2008; Kooy and Bakker, 2008). The selective extension of the water network was thus a way to create and emphasise difference, and on this basis to establish different privileges between European inhabitants of the colony and the indigenous population (Kooy and Bakker, 2008).

In many of these contexts, unequal infrastructure development produced in the colonial period persisted following decolonisation, where the state did not have access to sufficient capital to extend the water network to all its citizens (Gandy, 2008). Thus in cities water networks often developed in a patchwork manner, where well-off areas (often the former European quarters) were served by a well-developed network, while other zones were left dry (Gandy, 2008). The least well-served areas were generally the burgeoning informal neighbourhoods that had sprung up around cities (Bakker, 2003b), thus the exclusion from the water network was often associated to irregularity or lack of recognition from the state. In these areas, often the expansion of the formal water network had the effect of actually restricting access to water by rendering it contingent on connection to the network – for instance through the covering over of urban creeks or the prohibition of rain-water harvesting (Bakker, 2003b).

The importance of the citizenship/socio-economic rights/access to water nexus is particularly clear in the case of Durban, South Africa, during the Apartheid period. Despite the fact that infrastructure built in the first half of the 20th century had connected the white municipality and the adjacent Bantustan into the same network, a semiprivate water board was created to serve as intermediary between the municipal authorities and the administration of the Bantustan (Loftus, 2005). This was because Bantustans had been created as part of the policy of 'separate development' for members of different 'races'; if one of these territories had accessed water through a white municipality, this would have questioned the legitimacy and viability of the separatist project (Loftus, 2005). In this context, thus, the separation of water provision through the use of a private intermediary served to break the conceptual link between water provision and full South African citizenship (see also the chapters by Bond and Clark in this book).

Over the past decades, a common trend across different geographical settings is the increase of private sector participation in the water sector. While the specific ways in which this occurs are substantially different between the developed and developing world, there is a common ideological shift away from the conception of water as a socio-economic citizenship right. Privatisation of the water sector comes hand in hand with the conceptualisation of water as an economic good and users of the water system as consumers and claims that water network development should be based on full-cost recovery guaranteed by consumer payments.

It is important to note that these different paradigms of water resource development and regulation of access to water are not exclusive, and may exist simultaneously – articulated by different actors – in the same place. In fact, conflict between these different visions lies at the heart of many contemporary struggles over water resources. As we shall see, competing visions play an important role in the discourse on water in the contemporary West Bank.

The Palestinian case

Historic Palestine has a Mediterranean climate, in which the main water resources are the surface waters of the Jordan River and groundwater, of which the most important – and most contested – complex is the Mountain Aquifer which underlies the West Bank, and which may be divided into the Northern, Eastern, and Western Aquifer (see Figure 9.1).

Historical context

Under the British Mandate over Palestine (1917–1947), water resources were extensively monitored, regulated and drawn under centralised control (El-Eini, 2006). However, this centralisation was highly contested by the growing Zionist movement, which had implemented its own irrigation schemes and water institutions. Legislation was thus delayed and the operation of British Mandate institutions was often interrupted, resulting in a piecemeal development of water infrastructure. Overall, Arab agriculture in the region, which was generally extensive and based on surface water resources, benefited only marginally from British irrigation schemes and the Arab agricultural sector was generally hostile to British and Zionist initiatives, fearing a loss of historic water rights (El-Eini, 2006). Thus, already at this time, economic and social conflicts produced by the modernisation and centralisation of water resources were closely intertwined with conflicts between different groups with competing visions for the political and territorial future of the area.

The centralised water management initiated by the British Mandate was taken up by the Israeli state, after its establishment in 1948. This was accompanied by demographic shifts, as approximately 700,000 Palestinians had been driven from the area that had passed under Israeli control, alongside a substantial rise in Jewish immigration (in 1948 the Jewish population of Israel was approximately 1.5 million, Pappe, 2004). In order to support this immigration, large works of water infrastructure were created, such as the National Water Carrier, completed in 1964, which transported water from Lake Tyberias to the coastal cities and to the Negev desert with a capacity of 450 MCM per year (Mekorot, undated).

However, not all Israeli citizens benefited to the same extent from the development of water resources. The networks of many Arab villages suffered from poor infrastructure, and water quality problems persisted well into the

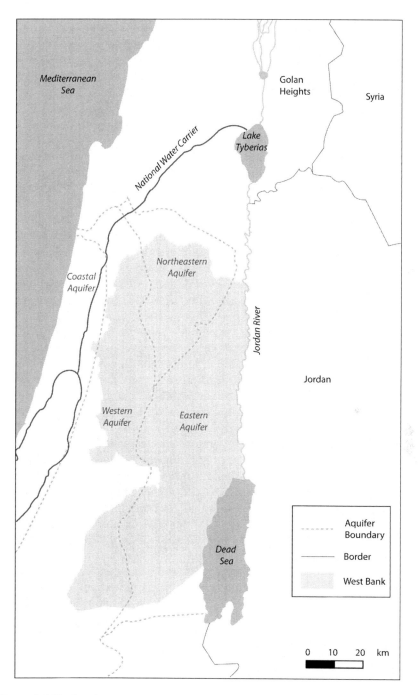

Figure 9.1 Regional water resources
Source: author

1990s (The Galilee Society, 2006). There was also a strong link between access to water and state recognition: the water network was not extended to about seventy unrecognised villages in the Galilee and the Negev desert, whose presence had not been recorded in official Israeli documents (Kanaaneh et al, 1995; The Galilee Society, 2006). In the agricultural sector, the system of water distribution placed the small-scale Arab-Palestinian landowners at a disadvantage in comparison to the collective holdings of kibbutz and moshav.[2]

Overall, during the first twenty years of Israel's existence, the state extended its control over water resources through its development of water legislation and large-scale water infrastructure. The development of water resources was a central pillar of nation-building, understood both as the physical development of the country and in terms of the configuration of the relationship between state and citizens. In this context, unequal access to water resources was an important element of differentiation between Palestinian-Arab and Jewish citizens of the Israeli state.

After the establishment of the state of Israel in 1948, the remainder of historic Palestine came under Jordanian (West Bank) and Egyptian (Gaza Strip) rule, and water resource development in these territories was incorporated within national resource planning of the two states. In the West Bank, both surface and groundwater resources were developed (Jayyousi, 2003; JICA, 2007), but in the agricultural sector no substantial changes in water allocation or management were carried out. In the domestic sector, regional water utilities were created and urban networks were developed (JWU, 1991). The channelling of water resources towards urban water provision initiated some tensions between urban and rural water users, as industrial water production for urban use threatened customary access to water. A particularly well known example of this was the conflict which occurred around the development of the Ein Samia well field for the provision of Ramallah, which substantially affected Ein Samia spring discharge, formerly a major source of water for agriculture and livestock for local peasants.[3]

Israeli occupation of the West Bank

In 1967, Israel occupied the West Bank and the Gaza Strip, and pursued a policy of economic and infrastructural integration (Brynen, 2000). Water resources of the territories were studied in detail by the Israeli national water resource development company Tahal in order to facilitate their incorporation within the Israeli system. On this basis, Israel pursued two simultaneous policies in the West Bank: limitation of Palestinian water resource development on one hand, and expansion of Israeli water infrastructure on the other.

In order to control Palestinian water resource development, Israel emended Jordanian water legislation through a series of military orders aimed at centralising authority in the hands of the Israeli civil administration. A few months after the 1967 occupation, military order 92 placed all powers previously vested in the Jordanian Government under Israeli control. As a consequence, all natural resources, including water, were directly administered by an Israeli officer. Military order 158, in turn, required that any development of Palestinian water infrastructure, including maintenance of wells, obtain a specific permit from the Israeli civil administration (Amnesty International, 2009). Permits were extremely difficult to obtain, and between 1967 and 1994 only 23 were accorded (World Bank, 2009).

At the same time, Israel developed its own infrastructure. Between 1967 and 1990, Israel dug 32 wells in the West Bank (Trottier, 1999). The development of the water network occurred in two different phases, linked to subsequent Israeli settlement policies of the territory. The first settlements were built in the Jordan Valley in the 1970s as part of the Allen plan, which sought to establish Israeli ongoing control on this border area with Jordan (Weizman, 2007). In this area an independent water network managed by the Israeli national water company Mekorot was developed, connecting the 25 deep Israeli wells to the settlements. During the second phase of Israeli settlement of the West Bank in the early 1980s (the Drobless plan), settlements were built throughout the territory (Segal and Weizman, 2003), and were connected to the Israeli national water network. In 1982, Mekorot also acquired ownership of all water infrastructure in the West Bank (Amnesty International, 2009).

The combination of limitations to Palestinian water resource development and the extension of the Israeli hydraulic system lead to increasing Palestinian dependency on the Israeli water system, both in the agricultural and in the domestic sector. In the agricultural sector, the development of Israeli wells had lowered the water table (in some areas up to twenty metres between 1976 and 2006, World Bank, 2009), reducing spring and well productivity, as well as water quality. In Al Auja, for example, an agricultural community north of the city of Jericho, a spring that previously had guaranteed over 9 MCM a year, had become a seasonal spring (PHG, 2004). The lowering of the water level together with the necessity to obtain Israeli permits to rehabilitate hydraulic infrastructure, reduced the number of Palestinian working wells from 184 in 1967 to 88 in 2006 (World Bank, 2009). Many communities were thus obliged to buy water from the Israeli national water company Mekorot. The economic burden this implied, combined with the denial of access to the Jordan River from 1967 onwards, forced some agricultural communities to settle elsewhere.[4]

The domestic sector also saw an increase of dependency on the Israeli system, both in villages and in cities. The extension of the Israeli network allowed for the connection of numerous Palestinian villages, financed by the Israeli Civil Administration or by private Palestinian or international capital.[5] In this way, many Palestinian communities began to purchase water from Mekorot, but the expansion of the network alone was not sufficient to guarantee improved access to water. In fact, communities that received water from Mekorot were often left dry, and the quality of networks was often sub-standard (Selby, 2003).

Following the Israeli occupation of the West Bank, Palestinian municipalities lost their ability to develop water resources and were obliged to purchase increasing amounts of water from Mekorot. The Jerusalem Water Undertaking, for example, which served the areas of Ramallah, Al Bireh and some parts of East Jerusalem, increased its purchase of water from 0.2 MCM a year in 1974 (approximately a fifth of what was produced from its own wells) to 5 MCM a year in 1994 (approximately two times what was produced from its own wells; JWU, 1995).

Overall, Israeli occupation of the West Bank halted Palestinian independent water resource development, and modernisation of water resources was carried out exclusively by the occupying power. This situation changed the meaning of conflicts around the production and distribution of water. Before the Israeli occupation, the first Palestinian attempts at hydraulic modernisation in the West Bank had caused internal tensions between rural and urban water users (as in the case of Ein Samia previously discussed). However, after the occupation these conflicts internal to Palestinian society were transformed into political conflicts between the occupying power and the occupied population. In this context, the struggle for access to water for Palestinians took on the significance of the struggle for a national collective right of self-determination over natural resources. This entailed two consequences: on the one hand it meant that internal Palestinian conflicts over access to water resources were perceived as secondary to the collective discrimination in access to water resources faced by West Bank Palestinians. On the other hand, the equation of water resource modernisation with the policy of the occupying power diverted attention from a broader discussion of the general merits of different modes of regulation of access to water resources.

The expansion of the water network also produced territorial consequences. As we have seen, many Palestinian villages were served through the Israeli water network, and some settlements were also connected to the network of the Palestinian-run Jerusalem Water Undertaking. Thus, the West Bank was integrated into the Israeli network to such an extent as to make subsequent territorial divisions extremely complicated. At the same time, the regulation of water fluxes enacted by this infrastructure contributed to the differentiation of the territory into Israeli zones inserted into modern fluxes of water, and Palestinian areas left partially or completely dry.

The West Bank after the Oslo Accords

The late 1980s witnessed substantial political change in the West Bank, as the First Intifada grew stronger in the Occupied Palestinian Territories. In 1993 the Oslo Accords were signed between the Palestine Liberation Organisation (PLO) and the state of Israel, and the Palestinian National

Authority (PA) was created with partial jurisdiction over the West Bank and the Gaza Strip. According to the Interim Accords, the PA obtained complete jurisdiction over infrastructure development in areas A and B (the Palestinian urban and peri-urban areas). In area C, which corresponded to approximately 60% of the West Bank territory, and which contained all Israeli settlements, infrastructure development remained under the jurisdiction of the Israeli Civil Administration (see Figure 9.2).

The creation of the PA was facilitated by substantial financial and logistical support from international development co-operation (Brynen, 2000), specifically international development organisations, United Nations institutions, the World Bank and many non-governmental organisations. The water sector was a key site of investment, attracting approximately three-quarters of a billion dollars between 1996 and 2004 (Zeitoun, 2008).

Thus, there were essentially three actors active in the Palestinian water sector – the Palestinian National Authority, the Israeli State and international development institutions – each of which had different approaches towards water resource development, informed by their different political understanding of the area, its history and its possible future. Let us examine more in detail their interaction and their impact on the Palestinian waterscape, which took place against a background of a general shift toward a commercial vision of water resource development, a vision partially adopted, but also partially contested, by each of them.

Legal and governance changes

The 1995 Interim Accords included an article specifically dedicated to the water resources in the West Bank (Article 40). This article formally recognised Palestinian water rights in the West Bank, although their precise definition was left to the Final Status Accords. Article 40 also divided the territory's underground water resources between Israelis and Palestinians according to existing patterns of use (thus conferring approximately 20% of the safe yield of the Mountain Aquifer to Palestinians and 80% to Israelis, see Table 9.1),

Table 9.1 Division of water resources according to the Interim Agreement

	Estimated safe yield	Israel	Palestinians
Western Aquifer	362 MCM/year	340 MCM/year	22 MCM/year
North-East Aquifer	145 MCM/year	103 MCM/year	42 MCM/year
Eastern Aquifer	172 MCM/year	40 MCM/year	54 MCM/year (+78 for future needs)
Mountain Aquifer	679 MCM/year	483 MCM/year	118 MCM/year (+78 for future needs)

Source: adapted from Amnesty International, 2009, p20

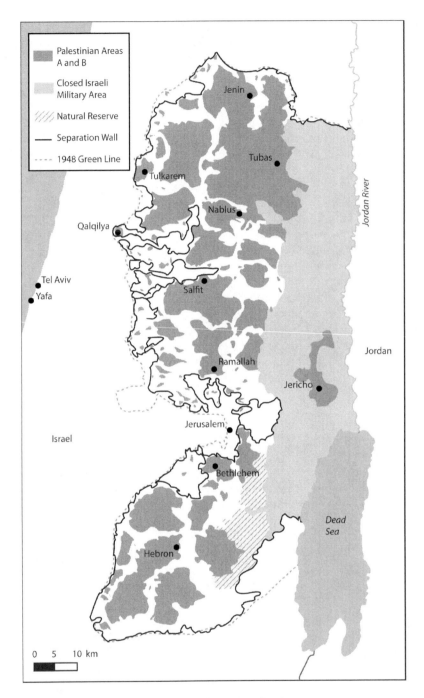

Figure 9.2 Post-Oslo territorial status of the West Bank
Source: author

and established that infrastructure serving only Palestinian localities in the West Bank would pass under Palestinian control (infrastructure serving both Palestinians and Israelis would remain under Mekorot control). However, the level of infrastructural integration that had occurred between 1967 and 1994 was such that only a portion of water infrastructure passed under Palestinian control (see Figure 9.3). At the same time, Israel began to construct an independent system for Israeli settlements that had previously been served from Palestinian sources (Israeli Water Authority, 2009).

Article 40 also established the Joint Water Committee (JWC), a joint body responsible for the construction and maintenance of water infrastructure in the West Bank. The JWC was made up of an equal number of Palestinian and Israeli members, but all decisions were made by consensus, and thus could be easily blocked. Also, JWC did not have jurisdiction within Israel's 1948 borders, even for projects that explored the shared Mountain Aquifer.

The Palestinian water sector also underwent a series of governance changes, broadly aimed at the centralisation and rationalisation of the system. Changes included the establishment of the Palestinian Water Authority (PWA) in 1996, and the implementation of the Palestinian Water Law in 2002. These changes were logistically and financially supported by international actors, particularly by the World Bank, which co-ordinated much multilateral donor support to the PA. The general economic framework followed by funders of the PA was based on the promotion of private sector initiatives, including the participation of the international private sector in the management of municipal water systems (Hall et al, 2002), rather than extensive support for the public sector. Also, funders strongly supported water regulation policies aimed at a commercialisation of the system, through policies such as full-cost recovery (PHG, 2004; World Bank, 2009).

Limitations to Palestinian water resource development

Despite the legal and institutional changes, and the logistical, technical and financial support of international organisations, in the years between 1995 and 2006, Palestinian water resource development witnessed very limited improvements. Several reasons explain this failure. A very significant one concerns the fragility and powerlessness of the PWA, as the PWA could operate only within narrow jurisdictional and legal limits. All decisions of the JWC had to be adopted by consensus and thus could be easily blocked or delayed. Only 57% of Palestinian projects were approved and these mainly concerned the recovery of urban networks, or the connection of new localities to existing networks (World Bank, 2009). Obtaining JWC approval for projects which would increase Palestinian water abstraction was particularly difficult and only 19% of proposed wells were approved, none of which for agricultural use.

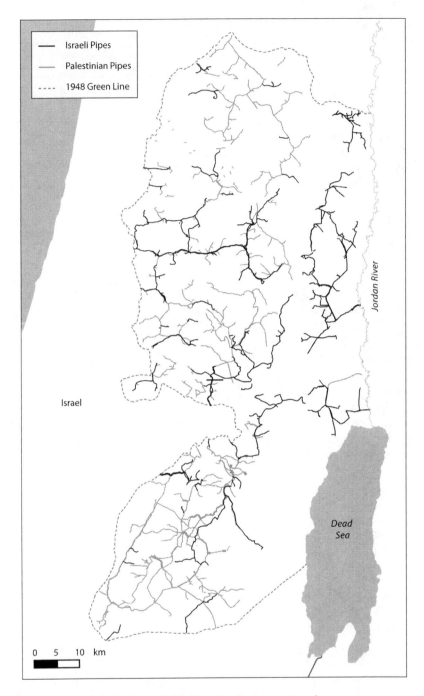

Figure 9.3 Post-Oslo division of the West Bank water network
Source: author

The territorial layout of the post-Oslo period placed further obstacles to Palestinian water development. Palestinian projects in area C had to be authorised by the Israeli Civil Administration as well as by the JWC (ARIJ, 2007); however, this authorisation was extremely difficult to obtain, thus preventing the development and upkeep of water infrastructure in this area (Amnesty International, 2009). This policy has been interpreted by various international organisations operating in the territory as part of a broader strategy to convince Palestinians to abandon area C (Amnesty International, 2009; COHRE, 2008). Furthermore, the creation of a closed military zone in all the Jordan Valley (see Figure 9.3), which limited access to 30% of the West Bank to Palestinians resident in the area, and the construction of the separation wall around the West Bank, which isolated at least 19 wells and 23 springs on the Israeli side of the wall,[6] prevented future Palestinian water resource development in two extremely favourable areas for ground-water abstraction.

A second factor limiting the action of the PWA was the modest level of its hydrological information. For instance, both the hydrological data and the estimates of Palestinian consumption used as a basis for the Interim Agreement negotiations had been provided by the Israeli hydrological services, due to the lack of data from the Palestinian side, and following Oslo it remained extremely difficult to obtain fundamental information such as detailed data on Israeli water abstraction in the West Bank.[7] A similar problem concerned the infrastructure transferred to the PA according to article 40: in many cases it had been passed to the PA without the necessary technical information to ensure its working and maintenance.[8]

Thirdly, the modernisation of the Palestinian water sector was also hindered by scarce financial resources. Certainly, immediately following the Oslo Accords, the water sector had been a key site of international investment. However, the difficulties encountered in carrying out projects, alongside the infrastructural destruction during the years of the Second Intifada – approximately $7 million worth of water infrastructure was damaged (Amnesty International, 2009) – made the international community increasingly reluctant to invest in the Palestinian water development. This was particularly marked in the private sector: the two international water management contracts in the Bethlehem and Gaza municipalities that included the participation of water multinationals Veolia and Suez were not renewed because of the unstable situation of the territories (Hall et al, 2002).

Overall, despite the creation of the PA and the economic and logistic support it received from the international community, changes to Palestinian water infrastructure were minor in the post-Oslo period. West Bank Palestinians continue to substantially depend on the Israeli water system: in the domestic realm, approximately 55% of domestic water in the West Bank is bought through the Israeli water system, reaching peaks of 80% in the area served by the Jerusalem Water Undertaking (World Bank, 2009).

Finally, the PWA also faced political difficulties because it attempted to be responsive to contradictory demands. Since international development organisations provided the major financial support for the PWA, this was pushed to adopt the neoliberal orientation promoted by much of the development community (World Bank, 2009). However, treating water as a commodity was in contrast with the principles embodied in the 2002 Water Law that considered water as a public good. Moreover, the PWA had also to deal with the discontent of some municipalities and village councils who considered the modernisation of the hydrological sector a threat to their customary regulation of access to water (Bossier, 2005 in Zeitoun, 2008). Thus, the PWA was obliged to juggle between contrasting positions, issued by different groups and organisations, trying neither to fully embrace nor fully reject any of them.

The public debate and the role of rights claims

The political, institutional and infrastructural changes produced in the West Bank in the post-Oslo period were accompanied by a broad local and international debate on water resources in the territory. In this debate, the three main actors involved in water resource development in the Occupied Palestinian Territories – Palestinian National Authority, Israeli State and International Organisations – articulated different visions of water resource development, largely connected to their broader visions for the geopolitical future of the territory. These three groups of actors were by no means internally homogeneous, nor were the individual actors consistent, as they sometimes adapted their framing of water issues according to the alliances they were seeking to forge.

Through this debate, it is possible to see how general concepts used in global discussions of water resource development and distribution acquired particular meanings when embedded in the Palestinian case. For all actors, a focus on questions of technical management of water resource development and distribution served to de-politicise a discussion of water resources which would bring up the political and territorial status quo of the West Bank. On the other hand, the notion of the right to water, even in its most generic conception, failed to garner consensus. Let us examine the debate in detail.

Palestinian actors

Palestinian actors articulated a variety of different discourses, some directly engaging with rights and others not. Parts of the PA, such as the PLO Negotiations Affairs Department, and various independent Palestinian organisations adopted a discourse of collective rights of self-determination over natural resources, and insisted on the recognition of these rights as a pre-condition for any discussion on the division of the region's water resources

(Zeitoun, 2008). At the same time, some Palestinian non-governmental organisations such as the Palestinian Hydrology Group framed the right to water according to a human rights framework (PHG, 2004), generally as part of a broader campaigning effort with international organisations.

In some instances, however, rights claims acquired a minor role. When dealing with international funders, for instance, the PA adopted a less politicised discourse centred on infrastructural and managerial modernisation of the water sector (Zeitoun, 2008). In this context, the PA also accepted the possibility of co-operating with the Israeli water sector regardless of their acceptance of Palestinian rights claims. Examples of co-operation include participation in the JWC or the signing of the Joint Declaration of Protection to keep water out of military operations during the Israeli re-occupation and siege of Palestinian cities during the Second Intifada (Zeitoun, 2008).

Finally, as previously mentioned, the 2002 Palestinian water law declared water a public good; however the Palestinian Authority expressed its openness to principles of commercialisation of the water sector, including the possibility of installing pre-paid water meters (Hanieh, 2008).

Israeli actors

On the Israeli side, two main visions of water resource management were at play, one state-centred and explicitly focused on military-territorial issues, and one more commercially oriented (Zeitoun, 2008; Alatout, 2010). The latter vision was prevalent in the international public debate, as it allowed Israel to dominate this debate through a language of expertise, technical efficiency and neoliberal rationality. Israel framed the regional water problem as one of scarce natural resources (Zeitoun, 2008) that could be solved through technical solutions in which Israel had a strong reputation. By presenting itself as the main regional expert in the use of water resources, the Israeli Water Authority could describe its position as 'technical' in opposition to a Palestinian 'ideological' stance:

> Israel believes that it can transform the issue of water from a possible source of controversy and tensions with the Palestinians to a basis for understanding and co-operation. In order to achieve this, the sides must dwell less on theoretical, legal or ideological aspects relating to the sharing of existing water sources, and focus more on practical and effective planning and preparation for co-ordinated water resources management . . .
>
> (Israeli Water Authority, 2009, pp2–3)

While the existence of Palestinian (domestic) water 'needs' was recognised by the commercially oriented parts of the Israeli water community (a principle which was not shared by the Israeli military-territorial perspective on water), these needs should be resolved through infrastructure development or sale of water through the Israeli system. Palestinian insistence on rights of self-determination over water resources, instead, was dismissed as obstructionist and idealist (Zeitoun, 2008). The generic notion of a Palestinian human right to water was also contested, possibly due to the fact that it was mobilised by the international community to criticise Israel's water policies in the OPT (Amnesty International, 2009).

International actors

International actors also held different positions, related in part to the different type of work that they carried out – technical co-operation or advocacy. Technical co-operation organisations tended to focus on the infrastructural and managerial modernisation of the Palestinian water sector, and to not engage in debates on the long-term territorial and political status quo of the West Bank. Thus, they were generally reluctant to take a position on a collective Palestinian right to water (Zeitoun, 2008). Also, while many organisations were critical of Israel's obstacles to Palestinian water resource development, they refrained from taking a strong position as they feared this would hinder their operations in the territory (Zeitoun, 2008).

When critiques of Israeli policies were voiced, the right to water was deployed by some actors, but not by others. On one hand, international advocacy organisations, such as the Centre on Housing Rights and Evictions and Amnesty International in their 2008 and 2009 reports respectively, addressed access to water in the Palestinian Territories within the framework of International Law and Economic, Social and Cultural Rights. In this case, the choice of framing the right to water as a *human* right was significant, as it allowed international organisations to address the issue of Palestinian access to water without engaging in the geopolitical status of the West Bank, and in the notion of a collective Palestinian right of self-determination over natural resources.

On the other hand, the language of rights (either of self-determination or human rights) was absent from the World Bank's 2009 'Assessment of restrictions on Palestinian Water Sector Development', published roughly at the same time as the Amnesty International and COHRE reports. This may be related to the contradiction between the notion of water as a human right and the commercial vision of water sector management promoted by the World Bank.

Conclusion

The Palestinian case represents a markedly different context from other contemporary struggles over access to water resources. In the West Bank, the perpetuation of selective and uneven access to water is not primarily produced by neoliberal restructuring of the water sector, but by on-going Israeli military and strategic interests in the territory. This entails that the focus of the internal Palestinian debate over justice and equity in access to natural resources mainly concentrates on issues of national sovereignty over natural resources. The conflation of conflicts over access to water resources with anti-colonial and self-determination struggles is not merely a contemporary affair; in fact, it has characterised struggles over access to water since Israeli occupation of the West Bank in 1967. Certainly, conflicts over the disruption of traditional patterns of access to water through the modernisation of the water system pre-date Israeli occupation of the territory. However, since following the occupation Israel became the sole agent of selective modernisation of the territory, aimed at supporting its own agricultural and domestic settlements, modernisation became essentially equated with a colonial enterprise.

This affected the political debate around the commodification of water resources. Certainly, water resources were treated as a commodity even before Israeli occupation, both in the domestic and in the agricultural sector. However, the modification of regional hydrology by Mekorot deep well abstraction and the extension of the Mekorot network throughout the territory of the West Bank meant that an increasing proportion of Palestinians were accessing water as a commodity delivered through the Israeli network. In this context, the general consequences of the 'enclosure of the hydro-commons' (Bakker 2003b, p338) in terms of equity in access to resources were less significant in the Palestinian public discourse than the implications of buying water produced through the colonial expropriation of Palestinian water resources.

The definition of the relation as a commercial one is important for the Israeli side. Israel presents its sale of water to Palestinians as a commercial relation between two states, despite the fact that the infrastructure that supplies Israeli settlements and Palestinian communities is essentially one system. Thus, in practice the same provider strongly discriminates in its water provision between settlers, full citizens of the state of Israel, and West Bank Palestinians. Although there are formally two state entities, Israeli control on territory and its resources is strong enough to justify the statement that in the West Bank, as in other colonial contexts and in the South African context, different levels of access to water reflect, reinforce and perpetuate different levels of citizenship amongst the inhabitants of the territory.

As we have seen in our discussion on Palestine and in our initial historical overview of the relation between water, rights and citizenship, the right to water acquires different meanings in different historical geographical contexts. Some of these include a basic socio-economic citizenship right (as was the case in state-planned water network expansion of mid-20th century Europe and North America), a collective right of self-determination over water resources (a particularly important vision in former-colonial contexts) and a universal human right (a framing which avoids engaging with the specific local politics of water resource development and distribution). All three of these visions were present in the debate over water resources in Palestine, and the choice of which one to adopt by different actors in the public debate was tied to their broader political position and to their strategic vision of the future of the territory. However, the adoption of *any* of these understandings of the right to water was refused by the military/ strategic parts of the Israeli water community, who generally refused Palestinian claims over resources, and, less explicitly, by Israeli and international actors who espoused a commercial vision of the water sector, according to which the definition of water as a right lay in contrast with its definition as a commodity.

Certainly, the strong politicisation of even abstract notions of the 'right to water' distinguishes the Palestinian context from other cases in which contestation occurs primarily around questions of implementation. However, our discussion of water politics in the West Bank underlines the importance of understanding the stakes at play in the deployment of rights discourses in different historical and geographical settings. This is a central question to be taken into account in debates around the effectiveness of the 'right to water' as a framework to advocate for justice in access to this resource.

Notes

1 This chapter is based on research carried out for my MA thesis at the University of Toronto. Field research for the thesis, involving interviews and the consultation of archival material, was carried out in Israel and in the Occupied Palestinian Territories between June and August 2009.
2 Interview with Senior Scientist, the Galilee Society, Shefa Amr, Israel, 17.08.2009.
3 Interview with Director of Operations, Jerusalem Water Undertaking, 08.08.2009.
4 Interview with Director of Major Palestinian NGO, Ramallah, Palestine, 14.08.2009.
5 Interview with Former General Director of West Bank Water Department, Jerusalem, 16.07.2009.
6 International NGO staff, personal communication, January 7, 2010.
7 Interview with Project Co-ordinator, Japan International Co-operation Agency, Ramallah, Palestine, 25.07.2009.
8 Interview with Former Director of West Bank Water Department, PWA consultant, JWC member, Jerusalem, 27.07.2009.

References

Alatout, S. (2010) 'Water multiples: settlement, territorial, and biopolitical assemblages in Israeli politics', unpublished manuscript discussed at the University of California Berkeley's Environmental Politics Workshop on November 12, 2010
Amnesty International (2009) *Troubled Waters – Palestinians Denied Fair Access to Water. Israel-Occupied Palestinian Territories*, Amnesty International, London

Applied Research Institute Jerusalem – ARIJ (2007) *Status of the Environment in the Occupied Palestinian Territories*, Applied Research Institute Jerusalem, Bethlehem, Palestinian Territories

Bakker, K. (2003a) *An Uncooperative Commodity: Privatizing Water in England and Wales*, Oxford University Press, Oxford

Bakker, K. (2003b) 'Archipelagos and networks: urbanization and water privatization in the South', *The Geographical Journal*, vol 169, no 4, pp328–341

Bakker, K. (2005) 'Neoliberalizing nature? Market environmentalism in water supply in England and Wales', *Annals of the Association of American Geographers*, vol 95, no 3, pp542–565

Brynen, R. (2000) *A Very Political Economy. Peacebuilding and Foreign Aid in the West Bank and Gaza*, United States Institute of Peace Press, Washington, DC

Centre on Housing Rights and Evictions – COHRE (2008) *Policies of Denial: Lack of Access to Water in the West Bank*, COHRE, Geneva

El-Eini, R. (2006) *Mandate Landscapes. British Imperial Rule in Palestine 1929–1948*, Routledge, London

The Galilee Society (2006) *Palestinian Arab Localities in Israel and their Local Authorities – A General Survey 2006*, The Galilee Society, Shefa Amr, Israel

Gandy, M. (2008) 'Landscapes of disaster: water, modernity and urban fragmentation in Mumbai', *Environment and Planning A*, vol 40, no 1, pp108–140

Hall, D., Bayliss, K. and Lobina, E. (2002) 'Water in the Middle East and North Africa (MENA) – trends in investment and privatisation', Public Service International Research Unit, www.psiru.org/reports/2002-10-W-Mena.doc, accessed 18.12.2009

Hanieh, A. (2008) 'Palestine in the Middle East: opposing neoliberalism and US power', *Monthly Review*, 19 July 2008, http://www.monthleyreview.org/mrzine/hanieh190708a.html, accessed 20.12.2009

Israeli Water Authority (2009) 'The issue of water between Israel and the Palestinians', http://siteresources.worldbank.org/INTWESTBANKGAZA/Resources/IsraelWaterAuthorityresponse.pdf

Japan International Cooperation Agency – JICA (2007) 'Feasibility study on water resource development and management in the Jordan River Rift Valley', unpublished document

Jayyousi, A. (2003) 'Water supply and demand development in Palestine: current status and future prospects', in Daibes, F. (ed) *Water in Palestine: Problems – Politics – Prospects*, Palestinian Academic Society for the Study of International Affairs (PASSIA), Jerusalem

Jerusalem Water Undertaking, Ramallah District – JWU (1991) *Annual Review 1991*, Jerusalem Water Undertaking, Ramallah, Palestinian Territories

Jerusalem Water Undertaking, Ramallah District – JWU (1995) *Performance and Prospects 1995*, Jerusalem Water Undertaking, Ramallah, Palestinian Territories

Kanaaneh, H., McKay, F. and Sims, E. (1995) 'A human rights approach for access to clean drinking water: a case study', *Health and Human Rights*, vol 1, no 2, pp190–204

Kooy, M. and Bakker, K. (2008) 'Technologies of government: constituting subjectivities, spaces and infrastructures in colonial and contemporary Jakarta', *International Journal of Urban and Regional Research*, vol 32, no 2, pp375–391

Loftus, A. (2005) 'A political ecology of water struggles in Durban, South Africa', DPhil Thesis, School of Geography and the Environment, University of Oxford, Oxford, United Kingdom

Mekorot (undated) http://www.mekorot.co.il/Eng/Mekorot/Pages/IsraelsWaterSupply System.aspx accessed 12.05.2011

Palestinian Hydrology Group – PHG (2004) *Water for Life. Israeli Assault on Palestinian Water, Sanitation and Hygiene during the Intifada*, Palestinian Hydrology Group, Ramallah, Palestinian Territories

Pappe, I. (2004) *A History of Modern Palestine*, Cambridge University Press, Cambridge

Segal, R. and Weizman, E. (2003) 'The mountain. Principles of building in heights', in Segal, R. and Weizman, E. (eds) *A Civilian Occupation: the Politics of Israeli Architecture*, Babel, London

Selby, J. (2003) *Water, Power and Politics in the Middle East: the Other Israeli-Palestinian Conflict*, Tauris, London

Simeone, T. (2009) Safe drinking water in First Nations communities. Parliamentary information and research service, Canada Library of Parliament, January 29, 2009

Trottier, J. (1999) *Hydropolitics in the West Bank and Gaza Strip*, Passia, Jerusalem

Weizman, E. (2007) *Hollow Land: Israel's Architecture of Occupation*, Verso, London

World Bank (2009) *Assessment of Restrictions on Palestinian Water Sector Development*, World Bank, Washington, DC

Zeitoun, M. (2008) *Power and Water in the Middle East. The Hidden Politics of the Palestinian-Israeli Water Conflict*, Tauris, London

10

WATER RIGHTS AND WRONGS

Illegality and informal use in
Mexico and the U.S.

Katharine Meehan

Introduction

'The rain barrel is the bong of the Colorado water garden.'
(Dave Phillips, columnist, *The Gazette*, 18 July 2007)

For nearly a decade, Colorado resident Kris Holstrom deliberately broke
the law every time it rained. 'Rain out here comes occasionally, and can
come very hard,' she explains (Riccardi, 2009), 'To be able to store it for
when you need it is really great.' Since 1999, Holstrom has harvested rain
and snowmelt from the peaked roof of her farmhouse. Tucked beneath
downspouts, metal drums capture and store runoff, which Holstrom uses to
irrigate her vegetable garden – a lush plot on a semi-arid plateau. But until
2009, Colorado outlawed rainwater harvesting because it impinged on downstream property rights, allocated to water users through a decades-old system
of 'first in time, first in right'.[1] Local opinions punctuated the debates between
water 'rights' and 'wrongs'. 'If you try to collect rainwater, well, that water
really belongs to someone else,' argues a member of the Colorado Water
Congress (Riccardi, 2009). 'I was so willing to go to jail for catching
water on my roof and watering my garden,' comments a Durango resident,
who harvests rain to feed his fruit trees (Johnson, 2009).

Thousands of kilometres away, in a poor neighbourhood of Mexico City,
Enrique Lomnitz adjusts the valve of a black plastic rain barrel – another
bootstrap reservoir operating outside the law. Designed to collect, filter, and
purify rooftop runoff, Lomnitz's system supplies the household with the
majority of its water. His organisation of activists, engineers, and local
residents has installed five systems in the surrounding community, with more
planned. Rainwater systems like this are 'replicable, economical, accessible,
and give good results,' Lomnitz explains (Godoy, 2009). Yet the rain in
these individual systems remains beyond the reach of Mexico's federal

159

authority, illustrating the potential friction between property rights and informal practices (Meehan, 2010). Whether in Colorado or Mexico, the rain barrel has become a lightning rod for political debate over water justice. Should water harvesters be considered 'criminals' or 'consumers'? What are the implications of so-called water 'wrongs' for conceptualising water 'rights'? What is the civic potential of civil disobedience for cultivating more humane and sustainable water futures?

In this chapter, I explore the potential of water illegalities to inform debates and struggles for the right to water. While scholars such as Karen Bakker (2007, 2010, and in this book) maintain a conceptual distinction between water rights (i.e., the allocation of property) and the right to water (i.e., as a moral claim, tactic, and strategy), I consider examples of informal water use that muddy such tidy analytics. As a starting point, I draw on the concept of 'popular illegalities' – the tacit and tolerated nonenforcement of laws (Foucault, 1977, 2000a, 2000b) – to trace shifting historical discourses of 'rights' and 'wrongs' within resource use. I then examine contemporary examples of water wrongs, drawing on examples of unauthorised water harvesting and reuse in the U.S. and Mexico. Such instances of illegality and informality, I argue, often blur the conceptual boundaries entrenched in rights talk and at times widen the aperture of political struggle for a right to water. This approach situates efforts for water justice across a spectrum of legal and social contexts, from affluent Colorado backyards to the patios of poor Mexican neighborhoods.

Why bother rethinking a right to water? While the moral compass of a human right to water is clear, critics point to the practical and conceptual problems embedded in rights talk. The current UN framework neither outlines a minimum quantity of water necessary for the 'full enjoyment of life' (UNGA, 2010); nor does it explicitly consider ecosystem needs, in-stream flows, or the rights of nonhumans (Bakker, 2010). It remains unclear which legal mechanism should hold governments accountable for implementation (Bakker, 2010), or how claims should be handled when rights are denied or violate other uses (Irujo, 2007). In contrast to the aims of anti-privatisation activists (e.g., Barlow, 2007), a right to water does not preclude private sector participation (Bakker, 2007, 2010, and in this book; Morgan, 2004, 2006a, 2006b). In fact, a rights mandate 'can be given technical and practical flesh via the entrepreneurial initiative of well-resourced private actors such as multinational water companies in combination with a regulatory framework controlled by public actors' (Morgan, 2004, p7). The dominant framework of rights talk hinges on abstract notions of universal entitlement and fails to account for variability in space, time, and legal tradition (chapters by Linton and Staddon et al; Morgan, 1999, 2004). 'Moreover,' argues Karen Bakker (2010, p218), 'the human right to water focuses only on drinking water and not on inherent aspects of rights to water resources: rights to water governance, rights to water resource use (complicated by their ties to land use and ownership), not to mention the contents (we might term this "institutional architecture") of these rights.' Clearly, the weaknesses of rights talk warrant new understandings.

In writing about illegality, I neither seek to condone unlawful or unethical activity, nor to champion civil disobedience as an automatic site of political struggle. Indeed, as Bronwen Morgan (1999, p316) warns, 'optimism *is* appropriate, but only in tension with pessimism. Both risk and opportunity inherently and necessarily reside in the core meaning of rights talk.' But alongside other chapters in this book, I use the tensions and contradictions of rights talk as a starting point. The next section considers the historical architecture of rights and wrongs to advance my argument.

From privilege to property: shifting constructs of illegality

Legal constructions of resource use 'rights' and 'wrongs', argues philosopher Michel Foucault (1977, 2000a, 2000b), are deeply influenced by historical power relations. During the period of aristocratic rule in feudal Europe, certain actions of civil disobedience – such as theft of wood, water, fodder, food – were widely tolerated by rulers and landowners, who saw such activities as necessary privileges for the survival of the working poor (Foucault, 1977; see also Hobsbawm, 1965, 1969; Linebaugh, 1976; Thompson, 1964). Such practices were 'popular' because of their diversity and prevalence.

> Sometimes it took on an absolutely statutory form – as with privileges accorded certain individuals and groups – which made it not so much an illegality as a regular exemption. Sometimes it took the form of a massive general non-observance, which meant that for decades, sometimes for centuries, ordinances could be published and constantly renewed without ever being implemented. Sometimes it was a matter of laws gradually falling into abeyance, then suddenly being reactivated; sometimes of silent consent on the part of the authorities, neglect, or quite simply the actual impossibility of imposing the law and apprehending the offenders.
>
> (Foucault, 1977, p82)

Illegality was deeply rooted in daily life. For law-makers as much as law-breakers, illegality was viewed as a tacit right to existence and necessary for the political and economic functioning of society (Foucault, 1977). Contrary to characterisations of theft as chaotic and disorderly, illicit activity often took a systematic form, subject to informal codes of conduct, internal hierarchies of power, and class-based expressions of political resistance (Foucault, 1977; Hobsbawm, 1965, 1969; Linebaugh, 1976; Scott, 1985; Thompson, 1964).

By the second half of the eighteenth century, the nature of illegality changed. In newly industrialised Europe, rapid urban growth and the accelerated circulation of wealth prompted new orderings of nature and society. Governments enclosed and parceled common lands. Elites redefined resource rights and wrongs. 'Landed property,' writes Foucault (1977, p85) 'became absolute property.'

> [All] the tolerated 'rights' that the peasantry had acquired or preserved (the abandonment of old obligations or the consolidation of irregular practices; the right of free pasture, wood-collecting, etc.) were now rejected by the new owners who regarded them quite simply as theft (thus leading, among the people, to a series of chain reactions of an increasingly illegal, or, if one prefers the term, criminal kind: breaches of close, the theft or killing of cattle, fires, assaults, murders *cf.* Festy and Agulhon). The illegality of rights, which often meant the survival of the most depraved, tended, with the new status of property, to become an illegality of property. It then had to be punished.
>
> (Foucault, 1977, p85)

As property, and with it access to natural resources, was reorganised under industrial capitalism, the guise of criminality also changed. Landlords no longer 'looked the other way' during the unauthorised collection of water or wood; resource theft was now punished under the regime of property.

Accompanying this shift, the punitive dimensions of popular illegalities were bifurcated into parallel systems of property-based and rights-based crimes. As Foucault (1977, p87) explains, 'This great redistribution of illegalities was even to be expressed through a specialisation of the legal circuits: for illegalities of property – for theft – there were ordinary courts and punishments; for the illegalities of rights – fraud, tax evasion, irregular commercial operations – special legal institutions applied with transactions, accommodations, reduced fines, etc. The bourgeoisie reserved to itself the fruitful domain of the illegality of rights.' Modern penal systems were not designed to *eliminate* crimes, but to *distinguish* particular 'rights' from 'wrongs' and assign punishment in differing ways. In short, what Foucault reveals are the subtle ways in which particular resources are divided into property rights and liberal rights, and their imbalanced mechanisms of enforcement and punishment.

Such distinctions are similarly invoked in contemporary water discourse and legal frameworks. On the one hand, though water is universally viewed as non-substitutable and necessary to life, it is often expressed as a liberal, universal *right* that cannot be denied. For example, in July 2010 the United Nations General Assembly enshrined the ethic of 'water for all' by unanimously passing a resolution on the 'Human Right to Water' (UNGA, 2010). 'There is now a widespread understanding that the right to water is an idea whose time has come,' insists Maude Barlow (2007, p196), 'These are not semantics: you cannot trade or sell a human right or deny it to someone on the basis of inability to pay.' Rights to water are thus constructed as *privilege*.

On the other hand, water is simultaneously constructed as *property*, and is thus subject to national and local legal traditions of *possession*. In Mexico, for example, the domestic catchment of rain eludes the very definition of 'water' as established by Article 27 of the Mexican constitution (Meehan, 2010). Rain is considered integral to the property it falls upon, unless that rainwater flows across property boundaries, becomes a stream (either permanent, intermittent, or torrential), or infiltrates aquifers and transforms into subsurface groundwater flows. Under this framework, harvested water precludes formal authority. But in the United States, the legality of water harvesting and reuse is far more variegated, due to its mix of common, civil, and even customary law traditions. Rainwater, for example, is governed on a state-by-state basis. In states such as Utah and Washington, collecting rooftop rainwater remains illegal, unless the harvester also owns water rights on the ground. In New Mexico and Arizona, however, municipalities have developed codes and regulations to actively encourage rainwater harvesting as a sustainable alternative to ground and surface water depletion. This paradox – water as privilege and possession – highlights the uneven legal geographies of harvesting and reuse.

Not surprisingly, citizens have bristled under such thorny legal ironies. 'The least favored strata of the population did not have, in principle, any privileges,' recounts Foucault (1977, p82), 'but they benefited, within the margins of what was imposed on them by law and custom, from a space of tolerance, gained by force or obstinacy; and this space was for them so indispensable a condition of their existence that they were often ready to rise up and defend it' (see also Scott, 1985). Though Foucault refuses to assume illegalities as automatic sites of resistance, he does locate struggle in the mundane, everyday practices of securing life and livelihood, even practices classified as 'wrong'.[2] From this angle, the 'practices of ordinary consumption can be moralised in ways that both challenge existing structures of authority and governance, and also draw powerfully on the collective agency of ordinary people' to imagine and enact different water futures (Morgan, 2008, p83). Water illegalities, as the next section suggests, reveal the contradictions and possibilities of enacting alternative struggles toward water justice.

Informality and illegality in the waterscape

Over one hundred kilometres outside of Mexico City, tucked in the rugged and poverty-stricken northwestern region of the state, residents of San Felipe del Progreso rely on water supply directly from the sky. An intricate rainwater harvesting system supplies the 6,000 town residents with potable water – meeting enough daily consumption needs that the community bottles and sells excess water on the regional market. The project has the capacity to supply 3.5 million litres of water each year, purified through an innovative filtration system and compliant with federal drinking water quality standards (Anaya Garduño and Martínez, 2007; Oldham, 2008). The installation was designed and built with support from Pro Zona Mazahua (a local development organisation) and the International Center for Demonstration and Training in Rainwater Harvesting (Centro Internacional de Demonstración y Capacitación en Aprovechamiento del Agua de Lluvia, or CIDECALLI), based at a nearby Mexican agricultural university. While CIDECALLI engineers helped design and build the systems, community members are responsible for running and managing day-to-day operations. For example, surplus profits from bottled water sales are reinvested in a community-managed fund. Local project leaders even invented the brand name 'Mazagua' to market the water, based on a combination of Spanish and Nahuatl words for 'rain' (Anaya Garduño and Martínez, 2007; Oldham, 2008).

Informal water use often transgresses or conflicts with legal boundaries, but in some cases water harvesters use this paradox to carve out new spaces of self-representation and control. When asked how the Mexican federal water commission (Comisión Nacional del Agua, or CONAGUA) reacts to the San Felipe project, one CIDECALLI engineer paused and chuckled:

> They [CONAGUA] are a bit jealous. We surprised them with our projects and our philosophy. At first, they stepped back and remained cool, disinterested. After all, they are in charge of supplying water to agriculture, livestock, industry, and finally to people. So of course they are looking for projects of a big scale, large volumes of water, lots of money. Rainwater harvesting doesn't interest them. [Pause] It's funny. We're the same people [engineers]. But we think differently about how to resolve water access problems, we think differently about water and people.[3]

CIDECALLI engineers, in this passage, use the extralegal space of rainwater harvesting to actively construct an alternative vision of water provision, distribution, and governance. As this participant in the San Felipe project observes, the biophysical quality of rainwater is not the only or even dominant issue – more significant is the tension between state authority and local control:

> I read a thesis by a student at the Colegio de México, who wrote about the implementation of a rainwater harvesting project in rural communities. She had done a survey of attitudes about rainwater. Actually, the majority of people thought rain was the safest form of water, above even piped [municipal] water and definitely above

bottled water. Just because they are wary of that outside influence, like the government.[4]

For those without the luxury of piped water, the idea of local control is also seen as a response, and potential pushback, to a chronic lack of infrastructure and resource provision (Bakker, 2010). For example, twin discourses of locality and informality are evident in the online description of Isla Urbana, a Mexico City-based water harvesting and urban design organisation:

> Isla Urbana trains plumbers in each neighborhood to install the rainwater harvesting systems. This way there is a pool of local knowledge to deal with any problems that may arise. We also support the local economy by purchasing all of our materials from the neighborhood hardware stores. Buying the materials locally also ensures that the communities have the resources necessary to maintain their systems. At Isla Urbana we believe access to water is a basic human right, and we are working to provide water to the citizens of the world one neighborhood at a time.
>
> (Isla Urbana, 2011)

Here, harvesters use 'local knowledge' and direct action – rather than mass mobilisation or protest – to cultivate new forms of water provision and authority. For some harvesters, questions of water informality and distribution also extend to ecosystems and nonhumans. Take the sentiments of this water harvester and activist from Tucson, Arizona (a rapidly growing desert city in the U.S. Southwest). She explains:

> Let's say we have a finite supply of water . . . conserving water just increases the carrying capacity for new houses. In fact, you are saving water for Tucson Water [the municipal utility]! There is not a single [government] agency that has the guts to quantify carrying capacity. The assured water supply program is partially trying to address that. But [the program manager] is always saying how the 100-year limit is a temporary band-aid.[5]

For this harvester, 'just' water distribution extends to people *and* the environment. Until such principles are enshrined in property rights (or the guts of government agencies), she advocates harvesting water while 'turning on the tap,' to drain municipal supply and thus force change among decision-makers:

> Many environmentalists refuse to conserve water, because they don't want to conserve water for future growth. Me, I harvest water *and* turn on the tap.

You know what would get them excited about water conservation? Take a riparian area that depends on a shallow aquifer, like the Tanque Verde. We pump groundwater there, so it is no longer a perennial reach. Since decreased water demand would increase supply, why not decrease groundwater pumping in that specific place and allow a permanent cut to the aquifer? The trick is, to maintain it over time. You'd have to ensure it in legal documents. But if you did that, I know tons of people who would conserve and harvest water. And I think that the water conservation efforts would dwarf any other strategies. You have to change public perception, but you have to give them a reason to change it.[6]

The combination of legal and illegal tactics may even enlarge the space of political participation and struggle (Morgan, 2006b, 2008). For example, the California-based collective Greywater Action (formerly known as the Greywater Guerrillas) uses renegade plumbing techniques not only to redirect household flows of water, but also as a form of protest:

My housemates and I discovered that with a hacksaw, some plumber's tape, and pipe cement, you can plumb and rebuild the foundation of your household's daily water consumption. It only takes a few minutes to detach your sink drain, a few hours to replace the drain with a tube filling your toilet tank with hand-washing water, and just a few days to design and build a greywater system, re-routing used shower water to irrigate a garden. After a few more days' work to replace the water-guzzling porcelain throne with a waterless dry or composting toilet, you are no longer so reliant upon dams and destructive water infrastructure.

(Woelfle-Erskine, Cole and Allen, 2007, p99)

Greywater reuse is the recycling of domestic wastewater (not including toilet effluent, known as 'blackwater') for non-potable uses. Examples of greywater systems include basic treatment wetlands, retrofitting sinks to flush toilets, dry composting toilets, and the reuse of laundry water with buckets. Such simple techniques and technologies are particularly common in less developed areas. But even among homes located *on* the municipal grid, at least in Mexico, urban dwellers recycle greywater to supplement water supply and subsidise household expenditures (Gaxiola Aldama, 2006; Meehan, 2010).[7] Greywater reuse often involves simple, accessible technologies and techniques, as evidenced by this woman in the Mexican city of Tijuana:

Doña Argélia lives in a densely populated canyon in western Tijuana. At least three times a week, she feeds laundry into her washing machine: a large-capacity machine bought second-hand from the United States. Her house is modest: concrete floors, walls made of particleboard, and a metal roof. Like her neighbors, Argélia's home has running water but no sewerage. Her recycling system is simple and pivots around her second-hand washing machine. First, she washes white laundry when the water is 'clear.' As the water 'darkens' after a load or two, she washes darker clothes. Finally, when the water is *sucia* [dirty], she uses it to irrigate plants and to keep down street dust in the dry season. She estimates that for every 6 loads of washing and rinsing cycles, at least 3 loads utilize recycled water.[8]

But like rainwater harvesting, greywater reuse often occupies an uncertain zone of legality. For example, until recently, redirecting even minor types of domestic wastewater violated public health and plumbing codes in the U.S. state of California. As Cleo, a member of Greywater Action, recounts:

> Laura and I had yet to crack a book on plumbing on greywater . . . [our] basement escapades reflected our plumbing naïveté. We bought some fittings and plumbed the bathtub drain into an old garden hose and ran it out to the garden. The greywater did get outside . . . two hours after a shower. Our roommates did not think we'd achieved greywater success. Laura's plumbing teacher – who was also the city code inspector – didn't think so either. 'You did what?!' he exclaimed when she asked him how to make the hose drain faster. The class was on appropriate drainpipe size. Draining a shower requires a two-inch pipe. Yes, code exists for a reason.
>
> (Woelfle-Erskine, Cole and Allen, 2007, pp140–141)

Code, in this context, refers to California's stringent residential plumbing regulations. Perhaps ironically, California's regulatory web outlawed (or made prohibitively expensive) most homemade greywater systems. Here, Laura of Greywater Action circumvents legalities in order to distribute water across a wider spectrum of uses and users:

> Sheepish but determined, we returned to the plumbing store and bought two-inch drainpipe and some more fittings, plumbed the shower through our homemade surge tank (a 55-gallon barrel laid sideways) and into an old bathtub out behind the house. We filled the bathtub with gravel and planted a few cattails into it. Our shower drained just like a normal shower, and it was great to send swirly greywater gurgling outside to water our plants.
>
> (Woelfle-Erskine, Cole and Allen, 2007, pp140–141)

But as Laura points out, doing the 'right' thing (i.e., enacting new forms of water distribution) uncomfortably rubs up against water rights. As she recalls:

> Now I was on a mission to learn everything I possibly could about water recycling. We'd already detached the bathroom sink drains and were bucket-flushing the toilet with the water. We had read about sinks that drain directly into the toilet tank. I wanted to try it, but when I surveyed the bathroom, it looked like gravity was against me. I decided to ask my plumbing teacher about it. The question didn't make it half way out of my mouth . . . 'Stop,' he said, 'Don't tell me any more of your crazy plumbing schemes.' The inspector in him had shut his ears to any interesting, innovative, and otherwise illegal plumbing scenarios.
> (Woelfle-Erskine, Cole and Allen, 2007, pp140–141)

Through her unauthorised plumbing schemes, Laura blurs the lines between greywater 'criminal' and crusading 'citizen' for water justice. In her Oakland backyard, water rights are directly cultivated in the practices of water wrongs.

Will harvesting rain and recycling greywater save the world? Perhaps not. But for some informal water users, such as these members of Greywater Action, water informality signals new modes of struggle:

> People like greywater treatment wetlands – bathtubs full of gravel and cattails – because they're one example of a human-scale answer to the problems of sewage treatment and mega-dams. If everyone built one in their backyard, would that bring the dams down? No. But guerrilla plumbers profiled in this book say that the taste of autonomy has gotten them to question the logic of the water grid.
> (Woelfle-Erskine, Cole and Allen, 2007, p6)

While the 'water grid' has colonised most imaginaries and aspects of water governance – from the failures to the ambitions – these cases of civil disobedience suggest that *informality*, and indeed even illegality, may be equally effective at realising and enacting a collective vision of 'water for all'.

The politics of civil disobedience

'I'm conflicted between what's right and what's legal. And I hate that.'
(Colorado resident, rainwater harvester, and
former U.S. National Park Ranger)[9]

Water informality, as the above quote reflects, frequently sits at the turbid confluence of provision and theft, privilege and property, rights and wrongs.

But such thick examples of civil disobedience also provide several clear insights for water justice and struggles. In the first place, water illegalities suggest that what is 'right' (e.g., the distribution of water in an equitable and sustainable manner) and what is 'lawful' (e.g., enforcing property boundaries) may be at odds – as the Colorado harvester above indicates. Following Foucault, I contend that this paradox is the outcome of modern conceptions of property and privilege, in which the legal right to own or possess water (i.e., property) is considered distinct from the liberal right to water (i.e., privilege). For Foucault (1977), this distinction springs from historical inequities and power relations, and is reflected in differing punishment regimes for property crimes and liberal rights. This asymmetry is evident in the water sector: systems for the allocation and enforcement of water rights are generally explicit and codified in law, while the current formulation of a human right to water is not equipped with the means to enforce universal service (Bakker, 2010). Consequently, formal tactics to ensure 'water for all' may be quite limited under what David Harvey (2010) calls our apparatus of [neo]liberal constitutionality, which promotes human rights while hollowing out rights to just conditions of production, consumption, and livelihood. In the meantime, water wrongs are, in part, trying to make it 'right'.

Water illegalities also trouble the conceptual binary of 'criminal' and 'consumer' embedded in rights talk. Through their everyday practices of illicit provision, harvesters in the U.S. and Mexico *both* supply a vital resource *and* break the law; they embody categories of *both* citizen *and* criminal. This dual postionality, argues Bronwen Morgan (2008, p82), illustrates how 'the boundary between citizen and consumer is an interactive, and fluid, product of the process of drawing two other boundaries: consumer/criminal, and consumer/subject of human rights. Civil disobedience in relation to consumption brings into sharp focus the salience of [these] two identities in constructing the limits of propriety in the zone of everyday ordinary consumption practices.' In other words, in advertently fulfilling the United Nations' explicit goal of 'water for all', the water criminal may simultaneously stand for human rights. For some, such as the residents of San Felipe and Mexico City, collecting rain is a necessary condition for survival but also a new mode of modern water organisation. For others, such as Cleo and Laura from Greywater Action, renegade plumbing is a rebel call for environmental sustainability and more sensible distribution. While not all citizens produce, access, or consume water in ethically just or humane ways, these vignettes do suggest that even 'criminal' actions may legitimately constitute a struggle for water justice.

In these cases, water illegalities also introduce a politics of distribution and commonality. Understandably, some scholars critique informality as an individuated mode of provision, thus undermining solidarity or collective efforts to achieve water justice (Bakker, 2010). But in contrast to *provision* – i.e., ensuring that houses are universally connected to the municipal grid – harvesting activities spotlight the politics of water *distribution*: the specific decisions, actions, and struggles involved in accumulating and allocating resources, a process that J.K. Gibson-Graham (2006, p98) argues is the ethical and fundamental basis of any commons. In both the U.S. and Mexico examples, harvesters accumulate water supply via individuated points of collection (shaped by the norms, values, and ethics of the household), yet they choose to distribute water to a variety of users (human and nonhuman) in a rather horizontal manner – creating, in effect, a commons. While the aggregate effect of these distributive practices remains an open question, their existence signals a different type of 'being-in-common'; one that offers different 'ethical coordinates' for 'negotiating interdependency' and commonality within water struggles, which may have the potential to foster connection among a plurality of movements (Gibson-Graham, 2006, p81).

At the same time, water informality is not a panacea of justice. Harvesting and reuse demands time, household [and typically gendered] labor, and can be risky for participants. Informal modes of water provision are not immune to the microgeographies of power, control, and exploitation – examples of water cartels and severe price inflation are common to many cities (Bakker, 2010; Gandy, 2006, 2008). Whether formal or informal, community management and collective action should be approached cautiously, as there is no guarantee that they are essentially more 'just' than state, corporate, or public-private models of water management (Bakker, 2008, 2010).

Yet, as these vignettes of illegalities indicate, water wrongs may also offer ethical coordinates for advancing redistributive water justice for people and the environment. If, as Karen Bakker (2010, p202) convincingly argues, we should 'turn our attention to the ways in which we can collectively act as stewards of the socioecological lifeworlds of which we are a part,' even acts of civil disobedience may be instructive for a right to water. While harvesting practices may be technically 'wrong', they are, in part, making it 'right'.

Conclusion

Drawing inspiration from Michel Foucault's writings on popular illegalities, this chapter has explored the civic potential of water wrongs in cases from the U.S. and Mexico. Though each site is embedded in unique material and socio-political contexts – underscoring the need to take seriously the geographical dimensions of legality and property (see, for example, chapters by Staddon et al and Mitchell) – I have shown how water wrongs transgress the conceptual boundaries of rights talk, underscore very real divisions between property and privilege, and in some cases even enlarge the space of political participation and struggle.

On a tactical level, these findings imply that universal 'rights' and local 'wrongs' will likely generate future political and legal friction in efforts to ensure 'water for all'. But simply legalising all forms of harvesting and reuse is not universally feasible – witness the tensions between rainwater harvesting and water rights precedence in Colorado, or the practical challenges of formalising an extralegal practice in Mexico. While it may be conceptually useful to separate 'property' from 'privilege' in struggles for the right to water (Bakker, 2007, 2010, and in this book), in practice they are more difficult to untangle, and indicate deep legal and social power asymmetries.

On the level of envisioning more humane and sustainable water futures, the rain barrels and renegade plumbing of harvesters like Kris Holstrom and Enrique Lomnitz are critical. Their distributive ethos, even in the face of enforcement, may offer an ethical blueprint for enacting more democratic surplus possibilities, particularly in the context of ongoing governance failure. Even such wrongs may prove vital to the struggle for rights.

Acknowledgements

I thank Peter Hoffmeister, Alex Loftus, and Farhana Sultana for helpful comments on earlier drafts. Part of this material is based upon work supported by the National Science Foundation under Grant No. 0727296, and fellowships from the National Estuarine Research Reserve System of the National Oceanic and Atmospheric Administration, Fulbright-Hays of the U.S. Department of Education, and the International Dissertation Research Fellowship Program of the Social Science Research Council with funds provided by the Andrew W. Mellon Foundation. Any opinions, findings, conclusions, and shortcomings expressed in this chapter are mine and do not necessarily reflect the views of the NSF, NOAA, USDE, or SSRC.

Notes

1 Colorado Senate Bill 09-080 allows limited collection and use of precipitation for residents who meet specific criteria. This legislation is a departure from Colorado's historically strict reading of the prior appropriation doctrine, the principle that guides water rights in much of the U.S. West. For details, see Colorado State University Extension (2011).
2 'There are many types of resistance,' explains Foucault, with typical evasive candor. His general refusal to identify absolute pathways of social change has been a point of frustration for many activists and critical theorists alike (see Harvey, 2007; Thrift, 2007).
3 Interview, 13 May 2008.
4 Interview, 9 May 2008.
5 Interview, 5 April 2006.
6 Interview, 5 April 2006.
7 For example, in a recent survey of 400 Tijuana households from a range of socioeconomic levels, nearly 26 percent of households recycled greywater for non-potable uses: such as for domestic cleaning, irrigation of plants and flowers, toilet flushing, and to keep down dust (Meehan, 2010).
8 Field notes, 27 August 2008.
9 Quoted in Riccardi (2009).

References

Anaya Garduño, A. and Martínez, J.J. (2007) *Sistemas de Captación y Aprovechamiento del Agua de Lluvia para uso Domestico y Consumo Humano en América Latina y el Caribe*, Colegio de Posgraduados en Ciencias Agrícolas, Texcoco, Mexico

Bakker, K. (2007) 'The "commons" versus the "commodity": Alter-globalization, anti-privatization and the human right to water in the global South', *Antipode*, vol 39, no 3, pp430–455

Bakker, K. (2008) 'The ambiguity of community: Debating alternatives to water supply privatization', *Water Alternatives*, vol 1, no 2, pp236–252

Bakker, K. (2010) *Privatizing Water: Governance Failure and the World's Urban Water Crisis*, Cornell University Press, Ithaca, NY

Bakker, K., Kooy, M., Shofiani, N.E. and Martijn, E.J. (2008) 'Governance failure: Rethinking the institutional dimensions of urban water supply to poor households', *World Development*, vol 36, no 10, pp1891–1915

Barlow, M. (2007) *Blue Covenant: The Global Water Crisis and the Coming Battle for the Right to Water*, McClelland and Stewart, Toronto

Colorado State University Extension. 'Graywater reuse and rainwater harvesting', www.ext.colostate.edu/pubs/natres/06702.html23, accessed 13 May 2011

Foucault, M. (1977) *Discipline and Punish: The Birth of the Prison*, Vintage Books, New York, NY

Foucault, M. (2000a) 'About the concept of the "dangerous individual"', in J.D. Faubion (ed) *Power*, The New Press, New York, NY

Foucault, M. (2000b) 'What is called "punishing"?', in J.D. Faubion (ed) *Power*, The New Press, New York, NY

Gandy, M. (2006) 'Planning, anti-planning and the infrastructure crisis facing metropolitan Lagos', *Urban Studies*, vol 43, no 2, pp71–96

Gandy, M. (2008) 'Landscapes of disaster: Water, modernity, and urban fragmentation in Mumbai', *Environment and Planning A*, vol 40, no 1, pp108–130

Gaxiola Aldama, R. (2006) 'Environment, poverty, and gender: Using and managing environmental resources in a Tijuana colonia', in J. Clough-Riquelme and N. Bringas Rábago (eds) *Equity and Sustainable Development: Reflections from the U.S.-Mexico Border*, Center for U.S.-Mexican Studies, La Jolla, CA

Gibson-Graham, J.K. (2006) *A Postcapitalist Politics*, University of Minnesota Press, Minneapolis, MN

Godoy, E. (2009) 'The goal: Not a drop wasted', *IPS News*, 19 September

Harvey, D. (2007) 'The Kantian roots of Foucault's dilemma', in J.W. Crampton and S. Elden (eds) *Space, Knowledge and Power: Foucault and Geography*, Ashgate Publishing, Hampshire, England

Harvey, D. (2010) 'Reading Marx's Capital with David Harvey', http://davidharvey.org/, accessed 15 May 2011

Hobsbawm, E.J. (1965) *Primitive Rebels: Studies in Archaic Forms of Social Movement in the 19th and 20th Centuries*, W. W. Norton, New York, NY

Hobsbawm, E.J. (1969) *Bandits*, Delacorte Press, New York, NY

Irujo, A.E. (2007) 'The right to water', *International Journal of Water Resources Development*, vol 23, no 2, pp267–283

Isla Urbana. 'What is Isla Urbana?', www.islaurbana.org/what_is.htm, accessed 15 May 2011

Johnson, K. (2009) 'It's now legal to catch a raindrop in Colorado', *The New York Times*, 29 June

Linebaugh, P. (1976) 'Karl Marx, the theft of wood, and working class composition: A contribution to the current debate', *Crime and Social Justice: Issues in Criminology*, vol 6, pp5–16

Meehan, K. (2010) *Greywater and the Grid: Explaining Informal Water Use in Tijuana*, PhD thesis, University of Arizona, Tucson, AZ

Morgan, B. (1999) 'Oh, reason not the need: Rights and other imperfect alternatives for those without voice', *Law and Social Inquiry*, vol 24, no 1, pp295–318

Morgan, B. (2004) 'The regulatory face of the human right to water', *Journal of Water Law*, vol 15, no 5, pp179–186

Morgan, B. (2006a) 'Turning off the tap: Urban water service delivery and the social construction of global administrative law', *European Journal of International Law*, vol 17, pp215–247

Morgan, B. (2006b) 'The North-South politics of necessity: Regulating for basic rights between national and international levels', *Journal of Consumer Policy*, vol 29, no 4, pp465–487

Morgan, B. (2008) 'Consuming without paying: stealing or campaigning? The civic implications of civil disobedience around access to water', in K. Soper and F. Trentmann (eds) *Citizenship and Consumption*, Palgrave Macmillan, Basingstoke, England

Oldham, F.L. (2008) 'Rainwater harvesting and rural development: The power of the CIDECALLI prototypes', www.harvesth2o.com/cidecalli_study.shtml, accessed 13 May 2011

Philipps, D. (2007) 'Logic seems absent from Colorado rainwater law', *The Gazette*, 18 July

Riccardi, N. (2009) 'Who owns Colorado's rainwater?', *Los Angeles Times*, 18 March

Scott, J.C. (1985) *Weapons of the Weak: Everyday Forms of Peasant Resistance*, Yale University Press, New Haven, CT

Thompson, E.P. (1964) *The Making of the English Working Class*, Pantheon Books, New York, NY

Thrift, N. (2007) 'Overcome by space; Reworking Foucault', in J.W. Crampton and S. Elden (eds) *Space, Knowledge and Power: Foucault and Geography*, Ashgate Publishing, Hampshire, England

United Nations General Assembly (2010) 'The human right to water and sanitation', document A/64/L.63/REV.1. 3 August 2010

Woelfle-Erskine, C., Cole, J.O. and Allen, L. (2007) *Dam Nation: Dispatches from the Water Underground*, Soft Skull Press, Brooklyn, NY

11

THE CENTRALITY OF COMMUNITY PARTICIPATION TO THE REALIZATION OF THE RIGHT TO WATER

The illustrative case of South Africa

Cristy Clark

While it has been widely recognized that the world is experiencing a water crisis, ongoing divisions remain in the quest to find an enduring resolution. The most polarized aspect of this debate relates to the role of market-based solutions and, particularly, the role of the private sector in the provision of water services. A related issue is the scope and content of the emerging human right to water and the question of whether or not it is compatible with corporate control over water.

This chapter argues that community participation plays a crucial – and insufficiently recognized – role in the realization of the human right to water. Community control over local water resources has a long history around the world and is the foundation of the modern concept of the 'water commons.' While some scholars have suggested that a 'commons approach' to water justice is a preferable alternative to the campaign for the human right to water, this chapter argues they are not only compatible, but mutually dependent. The recent South African *Mazibuko* water rights case (discussed below) highlights this issue and demonstrates the risk that the right to water will largely be a hollow right for poor communities unless it includes a complimentary right of community participation in water management.

The content of the right to water

Over the last decade there have been significant developments in the recognition of a human right to water (see introductory chapter to this book). However, there continues to be a lack of certainty over the status and content of the right to water, which has generated a predominately civil society-led

174

movement for an international treaty on the right to water (Friends of the Right to Water, 2005; Barlow, 2007). The fact that the World Bank, the World Water Council and many large private water corporations have officially accepted the existence of the right could be seen as an indication of the success of this campaign. However, these organizations have sought to reduce the content of the right to water to the question of *access* (Davidson-Harden et al, 2007, p30; Dubreuil, 2006; Salman and McInerney-Lankford, 2004). This approach is favoured because it is compatible with the introduction of market-based solutions to the water crisis, which proponents argue will increase efficiency and, thus, the total amount of water that is available (see, for example, Global Water Intelligence, 2010).

The problem with this argument is that free market water policies, including full cost recovery and private sector participation (PSP), do not necessarily improve efficiency or governance (Gutierrez et al, 2003). The commodification of water, in fact, often only serves to price it out of the reach of the poor (Gutierrez et al, 2003; Harvey, 2007; Bond, undated). For this reason the narrow definition has been the focus of intense scrutiny and critique, and many in the water justice movement have argued it is incompatible with the realization of the right to water (Harvey, 2007; Bakker, 2007 and chapter in this book; Barlow, 2007; Bond, undated). While much of this debate has, understandably, focused on the issues of affordability and financial sustainability, more attention now needs to be paid to the issue of community participation as economic considerations only represent a partial solution to the realization of the right to water, particularly for the most vulnerable.

The recent South African *Mazibuko* water rights case provides a clear lesson in how all of these issues intersect and highlights the risk that the right to water will be largely hollow if its content is reduced to the issue of access and it fails to incorporate the importance of community participation. The *Mazibuko* case is highly relevant to the development of any international approach to the right to water, as South Africa is one of only a few countries to constitutionally guarantee this right and the approach of its Constitutional Court to socioeconomic rights has been highly influential in the development of international jurisprudence.[1]

The *Mazibuko* water rights case

The impact of market-based policies on water access for the poor and marginalized was the focus of a constitutional challenge in South Africa in *Mazibuko and Others v City of Johannesburg and Others* CCT 39/09 [2009] ZACC 28. A group of Soweto residents from the suburb of Phiri challenged both the introduction of prepaid water meters into their community and the adequacy of the government-provided free basic water allocation. This case was the Constitutional Court's first opportunity to define the scope of the right to water, which is protected under section 27 of the South African Constitution. The applicants argued for an interpretation that provided a clear content to the right in terms of minimum quantities and the recognition of the need for affordability. They also sought to highlight some of the participatory elements of the right, including the need for procedural justice and community participation in the development and implementation of water policies.

These arguments were well received in the lower courts and provided some hope that an expansive interpretation of the right to water would be given judicial recognition in South Africa – a country that is widely viewed as a leader in social rights jurisprudence and which has come to serve as a model for their interpretation elsewhere. However, the Constitutional Court delivered a judgment that reflected a narrow interpretation of the right; one that not only accepted the dominance of market-based solutions, but also minimized the role of community participation in water management. This judgment threatens to make the right to water effectively meaningless for poor and marginalized communities in South Africa and, potentially, elsewhere.

Prepaid meters were introduced into Phiri under what Johannesburg Water called Operation *Gcin'amanzi* – an isiZulu word meaning 'conserve water.' One point of claim in the *Mazibuko* case was the applicants' contention the Phiri community had not been given sufficient opportunity to participate in the development of the policy and were not adequately consulted about its implementation.

Section 33 of the South African Constitution protects administrative justice,[2] and all administrative actions that affect the public must be preceded by public participation. This has been given legislative effect through the enactment of the *Promotion of Administrative Justice Act* 2000 (PAJA), which states, in section 4(1), any 'administrative action [that] materially and adversely affects the rights of the public' must be proceeded by a public inquiry or a notice and comment procedure. According to section 4(1) of the PAJA it is only permissible to depart from these requirements 'if it is reasonable and justifiable in the circumstances.' Section 4(2)(e) of the *Municipal Systems Act* (2000) also requires local Councils to consult the local community about the 'level, quality, range and impact of municipal services' and the 'available options' for the delivery of services.

The Constitutional Court dismissed the applicants' procedural justice-based arguments. In her majority judgment, Justice O'Regan found the decision to impose prepaid meters was an executive rather than an administrative action and, thus, did not need to comply with the PAJA.

South African human rights law professor Sandra Liebenberg (2010, postscript) argues this is an overly broad construction of the executive powers of a municipal council and that its 'effect is to reduce the scope for public participation in decisions which affect the enjoyment of constitutionally guaranteed socio-economic rights.' Liebenberg (2010, postscript) believes the decision to install prepaid meters constitutes 'policy formulation in a narrow rather than a broad sense' and, thus, constitutes an exercise of administrative rather than executive powers. This interpretation makes sense when you consider that the purpose of both Section 33 and the PAJA is to make government action more transparent and accountable. By classifying decisions that clearly impact on the rights of the public as executive rather than administrative the Court is effectively drawing a curtain around significant areas of public policy.

The Court also distinguished between the decision to introduce prepaid meters and the process of actually implementing that decision. Justice O'Regan held that this process of implementation had been procedurally fair, because meetings were held and community workers were employed to explain the process to the community. This finding contrasts starkly with the findings of Justice Tsoka who, in an earlier judgment for the High Court, described the City's consultative process as a 'publicity drive' for a policy decision that was already a *fait accompli* and found the City's approach towards the community to be 'misleading, intimidatory and presumptive' (*Mazibuko and Others v City of Johannesburg and Others*, Johannesburg High Court case no 06/13865 (as yet unreported), judgment of 30 April 2008). The Constitutional Court's decision to separate the decision-making process from the process of implementation allowed it to consider the community's participatory role without examining their power (or lack thereof) to affect the outcome of the process. A process in which community members are merely kept informed about the details of a decision that was made without their input is participatory only in the most passive sense of the word.

In 1969 Sherry Arnstein wrote a seminal article, *A ladder of citizen participation*, in which she set out eight levels of participation from nonparticipation, through tokenism, to citizen power. Arnstein (1969, p216) introduces this ladder by arguing, 'There is a critical difference between going through the empty ritual of participation and having the real power needed to affect the outcome of the process.' The City of Johannesburg's attempts at procedural justice fit squarely on the tokenistic rungs of Arnstein's ladder. This perspective was reflected in Justice Tsoka's judgment because he engaged in the kind of analysis of power that the Constitutional Court avoided.

A further participatory claim by the applicants related to the operation of the meters themselves. Under section 4(3) of the *Water Services Act* (1997) any disconnection or limitation of a water service must comply with a range of procedural safeguards, including the provision of 'reasonable notice' and an 'opportunity to make representations' to ensure that people are not denied access to basic water services for inability to pay.

The applicants contended that when a prepaid meter disconnects a water service for non-payment there is no notice and no opportunity to make representations. Instead they are forced to go without water until they are able to afford new credit or a new month's free basic water allowance becomes available. The uncontested evidence in the case was that most applicants' households were forced to go without water for an average of 15 days every month.

Despite this evidence the Constitutional Court found the safeguards in the *Water Services Act* could not have been intended to apply in these circumstances because it would be administratively impossible for a system with prepaid meters to comply with such requirements. This finding would seem to reinforce the applicants' contention that prepaid meters constitute an unlawful device, precisely because they are practically incompatible with the legislative guarantees of procedural fairness contained in the *Water Services Act*. However, the Court found instead that a water service is merely suspended rather than disconnected by the operation of the meters and that consumers are not denied basic water services because their free water allocation will still be available the following month.

The Court's finding that there is a tangible difference between the disconnection and the suspension of water services disregards the very real benefits afforded by the procedural safeguards. Not only do these safeguards give households a chance to prevent their water being disconnected, they also open a channel of communication that ensures that the government and water providers are not insulated from the knowledge that tariffs have been set at an unaffordable rate for many households. It is these benefits that make the opportunity to make representations enshrined in the Act an important expression of the right of participation.

This characterization of the operation of prepaid meters also disregards the impacts to health and dignity of having to endure the regular denial of water services due to lack of credit. Such a characterization is particularly surprising given this same issue was considered in the UK in *R v Director General of Water Services* (1998). The facts were strikingly similar, but in that case Justice Harrison found it was commonsense that prepaid meters disconnected 'or otherwise cut off the supply of water to the premises.' As a result, Justice Harrison found the use of prepaid meters to be illegal; agreeing with the concerns of the local authorities regarding the impact of automatic disconnection on vulnerable people on low incomes and on the flow on effect on public health. It is a harsh irony that the opposite conclusion could be reached eleven years later in South Africa – a country that is widely viewed as a leader in social rights jurisprudence.

At the conclusion of the *Mazibuko* case, the applicants argued that if their application were denied it would make cases such as theirs pointless and render the guarantee of social rights under the constitution hollow. Indeed similar concerns about the value of human rights to the struggle for water justice have been growing over the last few years (see Bakker, 2007 and chapter in this book; Bond's chapter in this book). This concern reflects a building frustration with the successful framing of PSP as compatible with the right to water and increasing cynicism over the co-option of human rights language by multinational water corporations.

In a recent article, Karen Bakker argued that instead of campaigning for a right to water, which Bakker describes as 'individualistic, anthropocentric, state-centric, and compatible with private sector provision of water supply,' the water justice movement should focus instead on reclaiming the global water commons (Bakker, 2007, p447). In the wake of the *Mazibuko* judgment, others, like Patrick Bond (see chapter in this book), have also began to ask whether Bakker is right to question the limitations of 'rights-talk' and to focus instead on reclaiming the commons.

In response to the applicants' concerns Justice O'Regan suggested this case should actually be seen as a victory for community participation in the dialogue over social rights in South Africa. She highlighted that the process of litigation itself led the City to reconsider some elements of its water services policies and to increase the amount of free water available to households on the indigent register.

Justice O'Regan is correct to point out that social rights litigation is a key expression of participatory rights for marginalized sections of the community. The capacity to litigate is a fundamental element of the participatory component of social rights and is central to their ultimate realization. Without the right to litigate it is very difficult to hold governments and non-state actors to account for violations. However, courts are not ideal venues for the poor and marginalized. As critical legal studies scholars have been pointing out for some time, litigation is expensive and time consuming, and rarely generates radical judgments on behalf of poor claimants (Langford, 2008; Pieterse, 2007; Bakan, 1997). Litigation may, therefore, be the wrong strategy to adopt, particularly if it involves diverting resources from activities viewed as more effective – such as public protests and political lobbying (Pieterse, 2007). There is also the risk of reinforcing the legitimacy of unjust laws due to the tendency of courts to uphold the status quo (Bakan, 1997). In this respect, scholars are right to be wary of the transformative potential of social rights litigation.

Bakker, for example, argues that a human right to water is compatible with the commodification of water (Bakker, 2007, p438; and chapter in this book). She is thus dismissive of 'rights talk,' arguing that it 'resuscitates a public/private binary that recognizes only two unequally satisfactory options – state or market control: twinned corporatist models from which communities are equally excluded' (Bakker, 2007, p440; see also Shiva, 2002). In place of this commodity-compatible 'rights talk' Bakker argues for a 'commons view of water,' which 'asserts its unique qualities: water is a flow resource essential for life and ecosystem health: non-substitutable and tightly bound to communities and ecosystems through the hydrological cycle' (Bakker, 2007, p441). These qualities mean, 'collective management of water by communities is not only preferable but also necessary' (Bakker, 2007, p441).

The commons view of water

Water has been collectively managed by many human societies throughout history (Boelens and Vos, 2006; Owley, 2004; Shiva, 2002, p24). Under Roman law, *aqua profluens* (flowing water) was classified as *res communes*, belonging to everyone in a way that emphasized equity and society wide ownership (UNDP, 2006). This *commons* view of water resources continues to be reflected in many systems of indigenous and customary water rights and in the community-participation that is fostered within the resulting water governance processes (Boelens and Vos, 2006; Owley, 2004; Shiva, 2002, p24).

This reality undermines one of the central arguments in Garrett Hardin's influential 1968 paper on *The tragedy of the commons*. Hardin (1968) argued that whenever a valuable resource is treated as part of the commons and not subject to private property rights, that resource would be quickly depleted, as there will be nothing to deter individuals from externalizing the negative consequences of overuse. While Hardin was right to point to the dangers of overuse and depletion, his conclusion that enclosure was the only viable solution fails to take into account the historical success of community management. The reality is that many local commons have historically been subject to 'well-defined rules of access and use' (UNDP, 2006, pp16–17). In relation to water, UNDP (2006, p185) points out, 'customary law often involves strict controls on water use, with water rights structured to balance claims based on inheritance, social need and sustainability. Institutional cooperation is common.'

Not only is cooperation at the community level common, it can also out-perform alternative models, such as central government management or market control, in terms of both efficiency and equity (Osmani, 2008, p14). One large-scale study into irrigation systems suggests that participatory community management is the 'superior institutional framework' in terms of ensuring both efficient and equitable water allocations (Ostrom and Gardner, 1993, p104; Ostrom et al, 1994). Similar findings have been made in relation to rural water supply projects and forest management, with community involvement increasing local ownership, willingness to contribute and concern for preservation (World Bank, 1994; Osmani, 2008, pp15–18).

The concept of the commons has also come to play a significant role in the global justice movement, representing an alternative to the market-centric model imposed by neoliberalism. However, while the commons may well be an alternative to neoliberalism and, as such, incompatible with the full scale privatization of natural resources like water, this does not mean that it is an alternative to human rights. Valid though they may be, critiques of human rights litigation must also take into account its strategic uses, including drawing attention to a campaign, starting a national dialogue on an issue and providing a vehicle for victims to speak out (see, for example,

Scheingold, 2004; Langford, 2008). It is worth noting, despite the unfavourable result, the applicants in the *Mazibuko* case and the social movement that sponsored their action do not regret their decision to litigate (Dugard, 2010). Instead they have pointed out that the litigation was not initiated in isolation, but was grounded in a much broader campaign against the application of neoliberal policies to the delivery of basic social services in South Africa (Dugard, 2010). The decision to litigate was made when their attempts at direct action through public protest and civil disobedience had been effectively crushed by the government through the use of private security forces and widespread arrests (Dugard, 2010; Groenewald, 2006).

Malcolm Langford points out, any 'critique of litigation as a vehicle for social change is only sustainable if there are viable alternatives or if litigation makes the situation worse in the absence of alternatives' (Langford, 2008, p42). This can be the case where litigation takes the place of, or is incompatible with, other strategies like mass community mobilization. However, litigation is often used as a last resort and in these situations such criticism is unsustainable.

It is, therefore, necessary to take a more nuanced approach to the value of litigation. As a mechanism for mobilization and progressive change it has its value. However, it is also necessary to acknowledge its limitations. One such limitation is the fact that litigation is only necessary when rights-protecting mechanisms have failed earlier down the line. Generally speaking, people will only litigate when there has already been a violation of their rights. This being the case any comprehensive protection of the right to water must also include mechanisms for protecting and realizing the right before policies are made and implemented, and this is where community participation is crucial.

By incorporating a participatory component, the right to water can respond to many of the concerns of those who seek to advocate for the global water commons. Calling the right to water 'individualistic, anthropocentric and state-centric' reflects a mischaracterization of the nature of social rights and contrasting it with the commons risks limiting the scope of this emerging right. Such an argument is based on a false dichotomy between the *commons* and a human right to water. Not only is it possible for the content of the right to water to incorporate much of the logic of the *commons*, this is in fact necessary for its full realization.

Participation

Rather than abandoning the campaign for a human right to water, the water justice movement should take advantage of the fact that the content of the right is still in the process of being negotiated and defined. There is scope to push for a definition that includes a clear participatory component and this would allow poor and marginalized communities the political space to demand inclusive and equitable water policies and services.

In General Comment No.15, the CESCR made it clear that the right to water contains a participatory component, particularly in relation to the regulation of water service providers (para 24) and the development of national water strategies (para 48). This fundamental requirement for public participation was also emphasized by the World Health Organization in their 2003 report on the right to water (WHO, 2003).

If the debate over the content of the right to water were to focus on participation, it would also help to reduce the current excessive focus on PSP. Concentrating so much attention on the relatively small number of private water providers risks missing the central issue: the structural barriers that prevent the poor and marginalized from accessing quality service from both public and private providers alike and their lack of power or 'entitlement' to demand better water policies (UNDP, 2006). This lack of power is equally relevant to publicly-managed systems, and community participation is the most promising mechanism for addressing this issue.

Of course, there are limitations and risks associated with participation. One of the main problems with participatory processes is that they are often based on fairly uncritical understandings of communities (Cooke and Kothari, 2001b, p6; Cleaver, 2001, pp44–46). Approaching communities as though they are homogenous units can serve to conceal power relations and the diversity of interests that exist within them (Cooke and Kothari, 2001b, p6; Cleaver, 2001). One of the big risks of such a naïve approach is that the most powerful groups within communities will dominate participatory processes and that these groups will be able to secure most of the benefits of the project for themselves (Gaventa, 2002, p5; Platteau, 2008, pp133–145). This is often described as the 'risk of elite capture' (Osmani, 2008, pp6–7; Cornwall and Gaventa, 2001, p16).

The second key criticism of participation is that it is often used as a mechanism for imposing pre-determined programs from above (Cooke and Kothari, 2001b, p5; Cornwall and Gaventa, 2001, p11). In these instances little devolution of power takes place despite the lip service paid to empowering marginalized communities. Even when participatory processes are not being deliberately used to co-opt communities, when they are applied in a formulaic way – or one that is focused more on efficiency than empowerment – they risk usurping more valuable locally-relevant processes or, at best, simply becoming meaningless (Cleaver, 2001; Aycrigg, 1998, p19).

Finally, participation is time and resource intensive, which can make it difficult to justify implementing (Aycrigg, 1998; Platteau, 2008, p128) and can create a barrier for women and poor households, who do not have sufficient time or resources to participate in projects or consultations (UNDP, 2006, pp10, 102, 194–195; Gaventa, 2002, p5).

Ultimately, however, these critiques and challenges do not negate the potential benefits of participation. Instead, they demonstrate that it is essential to ensure that participatory processes are properly grounded in a critical understanding of existing power dynamics; a commitment to empower-ment; and embedded in a broader process of improving governance (Guthrie, 2008, pp190–191). Increased social capital should increase the awareness of rights amongst disadvantaged groups and develop their capacity to claim them. Once this happens participation, empowerment and social capital can serve to reinforce each other (Osmani, 2008, p8). However, this positive cycle depends on the adoption of participatory techniques that go beyond information sharing and consultation. To be empowering, it is necessary for participation to actually shift power to the community (Cornwall and Gaventa, 2001).

This fully empowered community participation in relation to natural resources could also be described as 'commoning.' American historian Peter Linebaugh (2008, p297) argues that the commons can perhaps best be described as a verb or activity and that 'it expresses relationships in society that are inseparable from relations to nature.' It was through this process of 'commoning' that many historical rights were obtained. As economist Massimo De Angelis (2010, p2) argues, '[t]he important thing here is to stress that these rights were not "granted" by the sovereign, but that already-existing common customs were rather acknowledged as de facto rights.'

This 'commoning' or community participation can also provide a more accessible and responsive means of protecting and claiming rights than trad-itional legal remedies (Gaventa, 2002, p8). In recognition of these facts, both development and human rights theory increasingly view participation as a necessary 'prerequisite for making other rights claims,' and, as such, as a fundamental human right in and of itself (Gaventa, 2002, p3). This fram-ing of participation as a right in and of itself is also gaining increasing recognition within both human rights law and development practice, through the growing acceptance of the rights-based approach to development and of the central role that participation plays in strengthening good governance.

In keeping with this historical link to natural resource management, the strongest recognition of participatory rights exists within environmental management and water governance (International Law Association, 2004, art 18; Razzaque, 2009). Although this recognition has historically been fairly weak (mostly consisting of non-binding statements like Agenda 21, the Rio Declaration and the Dublin Statement on Water and Sustainable Develop-ment) in 1998 a right of public participation in environmental management was given far broader recognition through the United Nations Economic Commission for Europe *Convention on Access to Information, Public Par-ticipation in Decision-making and Access to Justice in Environmental Matters 1998* (the 'Aarhus Convention').

The Aarhus Convention has been described as signalling 'a shift from reactive participation to active participation at the local level and a collective manage-ment of shared water resources' (Razzaque, 2009). It has been influential in shaping the meaning of public participation within environmental management

183

and this shift in practice is starting to be reflected in numerous European Union Directives and in national legislation (Razzaque, 2009). This right of public participation has also been affirmed in relation to water governance through both General Comment No.15 and the Berlin Rules on Water Resources.

In the Berlin Rules, the International Law Association (ILA) argues a right of public participation in the management of water resources exists in customary international law. Article 4 declares a state duty to take steps to ensure public participation in the management of waters, while Article 18 emphasizes that people should be 'able to participate, directly or indirectly, in processes by which those decisions are made and have a reasonable opportunity to express their views on plans, programmes, projects, or activities relating to waters' (ILA, 2004). Article 18 also emphasizes the right of people to access information in order to participate in water governance, while Article 70 emphasizes the right to have effective administrative and judicial remedies (ILA, 2004).

The development of the Berlin Rules was significantly influenced by General Comment No.15, Agenda 21 and the Aarhus Convention 1998. In the commentary to Article 18, the ILA (2004) notes:

> In contemporary society, legitimacy largely depends on the consent of the governed, and hence on the sense that the governed have a voice through direct participation, representation, deliberation, or other methods. [. . .] Given the central importance of water in people's lives, the now generally recognized right of people to participate in decisions affecting their lives must apply to decisions concerning waters.

The ILA (2004) also notes that support for this proposition is provided in so many international instruments, 'that there can be little doubt that a right of public participation has now become a general rule of international law regarding environmental management even beyond the specific provisions of these agreements.'

It should be noted that the ILA did not accept the Berlin Rules unanimously. Four members of the ILA Water Resources Committee wrote a dissenting opinion in which they questioned whether these rules truly embodied current customary international law, or whether they were more appropriately described as expressions of emerging customary international law (Bogdanovic et al, 2004). This is a valid criticism. However, in relation to the content of the emerging international human right to water it is sufficient that these principles represent, at minimum, 'emerging customary international law.' Even that minimum status provides a strong foundation for the argument that a right of participation should be considered essential to the emerging content of the right to water.

Incorporating a participatory component into the right to water is an important mechanism for protecting the rights of the poor and vulnerable from the regressive impacts of market-based water policies. Even if the total amount of available water is increased by the introduction of these policies, without a right to water that is broad enough to include a participatory component, marginalized communities are unlikely to have the necessary power to challenge the structural barriers that prevent them from accessing sufficient water.

Renowned economist Amartya Sen expounded this distinction between the availability of a commodity and people's capacity to actually access it in his seminal study on poverty and famines (Sen, 1981). In this study Sen developed the concept of *entitlement* to explain the apparent paradox of famine or hunger in countries with food surpluses (Sen, 1981). According to Sen, '[p]eople suffer from hunger when they cannot establish their entitlement over an adequate amount of food' (Sen, 1999, p162). Sen's observation was that people tend to starve not for want of food, but for want of entitlement – that famine and starvation are related more to political and social arrangements than to the actual availability of food (Sen, 1999, p163).

This is why '[f]amines can occur even without any decline in food production or availability' (Sen, 1999, p163) and why genuinely protecting people from hunger has more to do with empowerment and human rights than it does with importing cheap food (Ziegler, 2004, p13). This recognition of the structural barriers that deny access to food for the vulnerable has led to a growing movement for the recognition that the genuine realization of the right to food can only be achieved if people have the 'right to define their own policies and strategies for the sustainable production, distribution and consumption of food . . .' (WFFS, 2001). This has been described as a demand for *food sovereignty* and has been championed by the world's largest social movement, *Via Campesina* (Ziegler, 2004, pp10–14; Via Campesina, 2001; Shiva, 2000, pp106–110).

The concept of food sovereignty challenges the more traditional approach to food security. Under a traditional food security approach the right to food has been interpreted to mean that people must have *access* to food (FAO, 2003).[3] While this is a laudable goal, this conception of food security has also been used to justify increased free trade and the accompanying increases in corporate control over global food production, through the assertion that these processes encourage economic growth and enhance the availability of cheap food (FAO, 2003, chp1). This approach ignores the fact that although trade liberalization may increase the availability of food, it often has a negative effect on equality, on people's control over food production and on export commodity markets, thus creating significant structural barriers to their capacity to actually *afford* sufficient food (Ziegler, 2004, p14).

The same kinds of structural barriers exist in relation to water and give rise to the same need for people to have the right to define 'their own policies and strategies for the sustainable *management*, distribution and consumption of *water*.' As UNDP (2006, p80) observes:

> The entitlements approach offers useful insights on water insecurity because it draws attention to the market structures, institutional rules and patterns of service provision that exclude the poor. It also highlights the underlying market structures that result in poor people paying far more for their water than the wealthy.

In a sense, arguing for the addition of a participatory component to the right to water is an argument for taking a *water sovereignty* approach to *water security*. While water security can be interpreted to mean the poor should have the right to *access* water (if they can afford to pay for it), taking a water sovereignty approach means that communities must also have the right to participate in the management of water resources. Under a water sovereignty approach, the right to water can be about so much more than the protection of consumer rights. This more expansive understanding of the right to water can and should incorporate much of the logic of the commons.

The potential outcome for the water justice movement of taking a water sovereignty approach to the human right to water has been highlighted by the recent referendums in Uruguay, Ecuador and Bolivia in which citizens voted for constitutional recognitions of a right to water that specifically guarantee community control over water services (COHRE, 2008).[4] As one Ecuadorean activist put it, 'At the moment we're still clients, water consumers; we pay our fee for water and we'll see you next month. Now we want to be *yakukamas* [keepers of water]' (Martines, 2009, p28). Empowering communities to be 'keepers of water' is likely to prove the most effective means of ensuring that water in the 21st Century is managed in a manner that is both sustainable and just.

Notes

1 For example, the incorporation of the concept of reasonableness into the text of Article 8(4) of the Optional Protocol to the ICESCR is a direct reference to the more limited form of judicial review adopted by the South African Constitutional Court in their seminal socioeconomic rights case, *South Africa v Grootboom*. This 'reasonableness review' was favoured by members of the Working Group and thus incorporated into the text of the Optional Protocol (see Porter, 2009, pp49–51, for a further discussion of this influence).
2 Constitution of the Republic of South Africa, Section 33.
3 See Food and Agriculture Organization of the United Nations (FAO) (2003) *Trade reforms and food security – conceptualizing the linkages*, at 24–31 for a full discussion of the wide variety of definitions that have been applied to 'food security' since the term was first used in the mid-1970s.
4 See *Constitucion de la Republica – Constitucion 1967 con las modificaciones plebiscitaras el 26 de novembre del 1989, el 26 novembre del 1994, el 8 de diciembre del 1996 y el 31 de octubre del 2004.*

References

Arnstein, S.R. (1969) 'A ladder of citizen participation,' *Journal of the American Institute of Planners*, vol 35, no 4, pp216–224

Aycrigg, M. (1998) 'Participation and the World Bank: success, constraints, and responses,' in *International Conference on Upscaling and Mainstreaming Participation: of Primary Stakeholders: Lessons Learned and Ways Forward*, Social Development – World Bank, Washington, DC

Bakan, J. (1997) *Just Words: Constitutional Rights and Social Wrongs*, University of Toronto Press, Toronto, Canada

Bakker, K. (2007) 'The "commons" versus the "commodity": alter-globalization, anti-privatization and the human right to water in the global South,' *Antipode*, vol 39, no 3, pp430–455

Barlow, M. (2007) *Blue Covenant: the Global Water Crisis and the Coming Battle for the Right to Water*, Black Inc, Melbourne, Australia

Boelens, R. and Vos, H.D. (2006) 'Water law and indigenous rights in the Andes,' *Cultural Survival Quarterly*, vol 29, no 4, pp18–21

Bogdanovic, S., Bourne, C., Burchi, S. and Wouters, R.P. (2004) *ILA Berlin Conference 2004 – Water Resources Committee Report Dissenting Opinion*, International Water Law Project, Berlin

Bond, P. (undated) 'The battle over water in South Africa,' http://www.africafiles. org/article.asp?ID=4564&ThisURL=./atissueforum.asp&URLName=AT%20 ISSUE%20FORUM, accessed 6 December 2005

Centre for Housing Rights and Evictions (COHRE) (2008) 'Ecuador's new constitution includes the right to water and sanitation,' http://www.cohre.org/ watsannews#article1119, accessed 14 June 2009

Cleaver, F. (2001) 'Institutions, agency and the limitations of participatory approaches to development,' in Cooke, B. and Kothari, U. (eds), *Participation: the New Tyranny?*, Zed Books, London and New York, NY

Constitution of the Republic of South Africa 1996

Cooke, B. and Kothari, U. (eds) (2001a) *Participation: the New Tyranny?*, Zed Books, London and New York, NY

Cooke, B. and Kothari, U. (2001b) 'The case for participation as tyranny,' in Cooke, B. and Kothari, U. (eds), *Participation: the New Tyranny?*, Zed Books, London and New York, NY

Cornwall, A. and Gaventa, J. (2001) 'From users and choosers to makers and shapers: repositioning participation in social policy,' *IDS Working Paper*, vol 127, Institute of Development Studies, University of Sussex, Brighton, UK

Davidson-Harden, A., Naidoo, A. and Harden, A. (2007) 'The geopolitics of the water justice movement,' *Peace Conflict & Development*, Issue 11, November

De Angelis, M. (2010) 'On the commons: a public interview with Massimo De Angelis and Stavros Stavrides,' *An Architektur, e-flux journal*, vol 17, June, http:// worker01.e-flux.com/pdf/article_150.pdf, accessed 15 February 2011

Dubreuil, C. (2006) *The Right to Water: from Concept to Implementation*, World Water Council, Marseilles, France

Dugard, J. (2010) 'Losing Mazibuko: (re) considering the campaign following judicial defeat,' Presentation at *Law's Locations: Textures of Legality in Developing and Transitional Societies*, University of Wisconsin Law School, 23–25 April (Quoted with permission from the author)

Food and Agriculture Organization of the United Nations (FAO) (2003) *The State of Food Insecurity in the World*, http://www.fao.org/monitoringprogress/index_en.html, accessed 12 October 2004

Friends of the Right to Water (2005) *Key Principles for an International Treaty on the Right to Water – Draft Work in Progress for Consultation*, http://www.blueplanetproject.net/documents/Key_Principles_Treaty_RTW_140405.pdf, accessed 30 January 2006

Gaventa, J. (2002) 'Exploring citizenship, participation and accountability,' *IDS Bulletin*, vol 33, no 2, pp1–11

Gaventa, J. (2004) 'Towards participatory governance: assessing the transformative possibilities,' in Hickey, S. and Mohan, G. (eds), *From Tyranny to Transformation*, Zed Books, London

Global Water Intelligence (2010) 'The human right to a national water plan,' http://www.globalwaterintel.com/insight/human-right-national-water-plan.html, accessed 20 August 2010

Groenewald, Y. (2006) 'Soweto starts its water war,' Mail & Guardian online, http://www.mg.co.za/article/ 2006-07-24-soweto-starts-its-water-war, accessed 15 May 2006

Guthrie, D.M. (2008) 'Strengthening the principle of participation in practice for the achievement of the Millennium Development Goals,' in United Nations Department of Economic and Social Affairs (ed) *Participatory Governance and the Millennium Development Goals (MDGs)*, United Nations, New York, pp163–191

Gutierrez, E., Calaguas, B., Green, J. and Roafand, V. (2003) *New Rules, New Roles: Does PSP Benefit the Poor? – Synthesis Report*, WaterAid and Tearfund, London

Hardin, G. (1968) 'The tragedy of the commons,' *Science*, vol 62, pp1243–1248

Harvey, E. (2007) 'The commodification of water in Soweto and its implications for social justice,' Thesis submitted for doctor of philosophy, Sociology Department, School of Social Sciences, University of the Witwatersrand, Johannesburg, South Africa

International Law Association (2004) *The Berlin Rules on Water Resources*, Fourth Report, Berlin Conference on Water Resource Law, Berlin, Germany

Langford, M. (2008) 'The justiciability of social rights: from practice to theory,' in Langford, M. (ed), *Social Rights Jurisprudence: Emerging Trends in International and Comparative Law*, Cambridge University Press, New York, NY, pp3–45

Liebenberg, S. (2010) *Socio-economic Rights: Adjudication under a Transformative Constitution*, Juta Academic, Cape Town, South Africa

Linebaugh, P. (2008) *The Magna Carta Manifesto: Liberties and Commons for All*, University of California Press, Berkeley

Martines, J.P. (2009) 'Keepers of water,' in Bell, B., Conant, J., Olivera, M., Pinkstaff, C. and Terhorst, P. (eds), *Changing the Flow: Water Movements in Latin America*, Food and Water Watch, Other Worlds, Reclaiming Public Water, Red VIDA and Transnational Institute, pp27–28

Mazibuko and Others v City of Johannesburg and Others (2008) All SA ZAGPHC 128

Osmani, S. (2008) 'Participatory governance: an overview of issues and evidence,' in United Nations Department of Economic and Social Affairs (UNDESA) (ed), *Participatory Governance and the Millennium Development Goals (MDGs)*, United Nations, New York, NY, pp1–48

Ostrom, E., Lam, W. and Lee, M. (1994) 'The performance of self-governing irrigation systems in Nepal,' *Human Systems Management*, vol 13, no 3, pp197–207

Ostrom, E. and Gardner, R. (1993) 'Coping with asymmetries in the commons: self-governing irrigation systems can work,' *Journal of Economic Perspectives*, vol 7, no 4, pp93–112

Owley, J. (2004) 'Tribal sovereignty over water quality,' *Journal of Land Use and Environmental Law*, Fall

Pieterse, M. (2007) 'Eating socioeconomic rights: the usefulness of rights talk in alleviating social hardship revisited,' *Human Rights Quarterly*, vol 29, pp796–822

Platteau, J.P. (2008) 'Pitfalls of participatory development,' in United Nations Department of Economic and Social Affairs (ed), *Participatory Governance and the Millennium Development Goals (MDGs)*, United Nations, New York, NY, pp127–159

Porter, B. (2009) 'The reasonableness of Article 8(4) – adjudicating claims from the margins,' *Nordisk Tidsskrift For Mennekerettigheter*, vol 27, no 1, pp39–53

R v Director General of Water Services (1998) (Queen's Bench Division, Crown Office List)

Razzaque, J. (2009) 'Public participation in water governance,' in Dellapenna, J. and Gupta, J. (eds), *The Evolution of the Law and Politics of Water*, Springer, Netherlands

Salman, S.M.A. and McInerney-Lankford, S. (2004) *The Human Right to Water: Legal and Policy Dimensions*, World Bank, Washington, DC

Scheingold, S. (2004) *The Politics of Rights: Lawyers, Public Policy and Social Change*, University of Michigan Press, Ann Arbor, MI

Sen, A. (1981) *Poverty and Famines*, Oxford University Press, New York, NY

Sen, A. (1999) *Development As Freedom*, Oxford University Press, Oxford

Shiva, V. (2000) *Tomorrow's Biodiversity*, Thames and Hudson, London and New York, NY

Shiva, V. (2002) *Water Wars*, South End Press, London

Stiglitz, J. (2002) *Globalization and its Discontents*, Allen Lane, London

United Nations Committee on Economic, Social and Cultural Rights (CESCR) (2002) *General Comment No. 15 on the Right to Water*, E/C.12/2002 /11

United Nations Development Programme (UNDP) (2006) *Human Development Report – Beyond Scarcity: Power, Poverty and the Global Water Crisis*, UNDP, New York, NY

Via Campesina (2001) *Priority to People's Food Sovereignty*

World Bank (1994) *World Development Report: Infrastructure for Development*, Oxford University Press, New York, NY

World Forum on Food Sovereignty (2001) *Final Declaration of World Forum on Food Sovereignty*, http://www.ukabc.org/havanadeclaration.pdf, accessed 12 June 2010

World Health Organization (WHO) (2003) *The Right to Water*, http://www.who.int/entity/water_sanitation_health/rtwrev.pdf, accessed 10 March 2004

Ziegler, J. (2004) *Trade and Food Security: the Failure at Cancún, Food Sovereignty and the Right to Food, Transnational Corporations and the Right to Food*, Report submitted by the Special Rapporteur on the right to food, E/CN.4/2004 /10

12

THE RIGHT TO THE CITY AND THE ECO-SOCIAL COMMONING OF WATER

Discursive and political lessons from
South Africa

Patrick Bond

Introduction

To genuinely contribute to a 'right to the city,' a crucial challenge for water rights advocacy is to transcend narrow juristic narratives that, as Karen Bakker (2007, p447, and chapter in this book) argues, tend to be 'individualistic, anthropocentric, state-centric, and compatible with private sector provision of water supply.' This challenge became acute in South Africa on 8 October 2009, when in *Mazibuko v Johannesburg Water*, the Constitutional Court overturned two lower-court rulings that had earlier been celebrated by the urban social movements of Soweto and comparable organizations across South Africa and the world, as well as academics (Bond and Dugard 2008 and *Mazibuko & Others v the City of Johannesburg & Others*, 2008). That court case provides the basis for rethinking both rights and commons so that both the ecological and the community-control factors are foregrounded, alongside contestation of the deeper logic of capital accumulation that explains the drive to water commodification within which activists campaign for water rights.

Some such campaigns win, but others lose. The *Mazibuko* case revolved around the amount of water each person needed (on average) each day – the Soweto plaintiffs demanded 50 liters, and the City insisted that 25 was sufficient – and whether the water would be delivered through ordinary credit meters (as the plaintiffs demanded), or on a pre-payment basis (as Johannesburg was doing in low-income areas). The latter was argued to be in violation of the Constitution, anti-discriminatory provisions in the Water Services Act and the Johannesburg water by-laws. This case was the most important test, so far, of possibly the world's most advanced water rights Constitutional clause: 'everyone has the right to an environment that is not harmful to their health or well-being . . . everyone has the right to have access to . . . sufficient water' (Republic of South Africa, 1996). The strongest hopes for *Mazibuko* were expressed by the talented lawyer who was central to developing strategy on the case, Jackie Dugard (2010a):

> Rights can be useful to the left, regardless of the ultimate outcome of litigation per se. Advocating a pragmatic approach to rights, I suggest that in contemporary South Africa, with its extreme socio-economic and racial inequalities, while in the normal course of events the law does indeed serve the interests of elites, rights-based legal mobilisation can have a predominantly positive impact on social movements representing disempowered groups, including the poor . . . . If strategically used, rights-based legal mobilisation may in certain circumstances offer the left an additional tactic in a broader political struggle. In some instances the additional tactic might be a last resort, but it remains a useful one.

These words preceded the October 2009 defeat (for Dugard and colleagues' immediate reaction, see Centre for Applied Legal Studies, 2009 and Coalition Against Water Privatization, 2009). The benefits and costs of litigating *Mazibuko* require full and frank debate, and only then can the political lessons for the broader 'right to the city movement' be elaborated. For example, in her review of *Mazibuko*, Cristy Clark (in this book) insists that the right to water will lack any substance for poor communities unless it includes a complementary right to participation. That is one lesson, but unless capable of breaking beyond the bounds of neoliberal public policy, it is one acceptable to many water privatizers (as pointed out by Bakker 2007 and in this book) and other neoliberal advocates of a smaller state.

Moreover, as Chad Staddon, Thomas Appleby and Evadne Grant (in this book) argue, it was a further blow to the plaintiffs when the Court ruled that it was inappropriate for Courts to get involved in setting prescriptive levels of water provision. This suggests that putting this kind of public policy power in the hands of (quite conservative) judges may be inappropriate. Even more generally, Staddon et al suggest, there is a huge paradox in the fact that complainants are often forced to appeal to the very states (often either failed or authoritarian) who have failed to provide them with water.

In addition to these political concerns, the full set of hydropolitical connections between the social and ecological are often not properly con-ceptualized within a water rights framing. The politics of water rights are, therefore, better contextualized from a Marxist right to the city stand-point advocated by Lefebvre and updated by Harvey, given that both political-economic and political-ecological concepts can be deployed within the strategies of urban social movements. This we see initially by way of understanding the limitations to the *Mazibuko* case.

South Africa's dash of cold water on the right to water

There were, in retrospect, many negative lessons about *Mazibuko*. This foundational case could be criticized on grounds it was:

- individualist: private/familial instead of public/political
- consumption-oriented, without linkages to production and ecology
- framed not to resist but to legitimize neoliberalism
- unable to transcend society's class structure, and thus in the process it distracted activists from potentially more serious strategies to dismantle class divisions through redistribution and reparations
- technicist, thus alienating the mass base and society in general
- guilty of making mass-based organizations the 'client' which in the process became 'domesticated' (Madlingozi, 2007), for example, told to halt protests during litigation
- subject to the 'watering down' of rights, given SA Constitutional clauses of 'progressive realization,' and of 'reasonable' measures 'within available resources'
- tempting for scholar-activists to follow its legal alleyways, which in turn distracted from a more transformative route to politics
- dangerous in class-power terms, insofar as judges are amongst society's most conservative elites
- reflective of the overall problem that even liberal-democratic capitalism won't deliver basic-needs goods to poor people.

Elsewhere (Bond, 2010) I have delved into these specific problems in more detail, as did a group of critical legal scholars in more general terms, debating whether rights narratives are optimal for progressive South African politics: Danie Brand (2005), Tshepo Madlingozi (2007), Marius Pieterse (2007) and especially Daria Roithmayr (2011). Since Clark (in this book) provides detailed case analysis (see also Danchin, 2010), it is worthwhile to follow through the political implications in this contribution, including the relationship of the South African water struggles to the right to the city.

One mistake was the narrowness of the litigant's request for relief partially on grounds of international evidence of minimal water needs, because according to Peter Danchin (2010), the Constitutional Court:

> signaled that while international law is relevant and helpful for constitutional analysis (as the Constitution itself requires) it does not intend to adopt the minimum core approach but rather will develop its more flexible reasonableness doctrine in an effort to forge

a distinctly South African attitude to the justiciability of economic and social rights.

But as Dugard (2010b) replied, the way 'reasonableness' was posed ignored the *realpolitik* of Soweto:

> First, the Constitutional Court misunderstood the applicants as arguing for a minimum core approach to the right to water. They did not. Rather, the applicants pursued the approach established by the Constitutional Court in *Grootboom* (in its rejection of the minimum core content approach, as being too inflexible), which is that rights and obligations can only be established in context. This is precisely what the applicants did in *Mazibuko*: they asked the Court to determine the reasonableness of the City's Free Basic Water policy in the context of a high-density urban township with waterborne sanitation and no alternative water or sanitation sources. The Court, however, cast this as a minimum core content argument. And, displaying an extraordinary degree of deference, found the City's Free Basic Water policy to 'fall within the bounds of reasonableness,' which appears to me to be a worrying retreat from the standard of reasonableness and of inquiry set in *Grootboom*.

However, in rebuttal to Dugard it might be argued that the *Mazibuko* plaintiffs (and especially their legal team) did not stress strongly enough the extent to which wealthy white residents had access to plentiful, inexpensive water on credit (not pre-paid), for comparative water consumption across race and class was not a major part of the case, as the effort to win a victory meant narrowing the narrative to a relatively non-contextualized terrain.

Dugard (2010b) then points out other areas where the Constitutional Court justices appeared both class- and race-biased:

> Second, contrary to the findings of both the High Court and the Supreme Court of Appeal, the Con Court found the City's interpretation of the by-laws as allowing the installation of prepayment meters to be 'textually permissible', which seems to be a new form of highly deferent legal interpretation. Third, in dismissing the applicants' arguments that prepayment meters amount to unfair discrimination based on race – because, despite proven debt across the City, prepayment meters have only been installed in poor black areas – the Court said that the applicants had not proven that prepayment meters were installed in ALL black areas. This is nonsensical and goes against all its previous equality decisions. It would mean that, for example, if I allege that my dismissal on the grounds of my sexual orientation (a listed ground in the Constitution)

amounted to unfair discrimination, I would have to prove that my employer had dismissed all other gay employees in the organisation. In South Africa, there is growing concern about the *Mazibuko* judgment and the Court's apparent retreat from enforcing socio-economic rights.

The legal-technicist arguments deployed by the Constitutional Court were thus subtly political, in defense of the status quo. These arguments would lead not only to denial of water to low-income people, but also to a confirmation of segregatory processes in South Africa's cities. It is here that we see the broader merits of a 'right to the city' campaign that avoids *Mazibuko's* pitfalls, through awareness of the simultaneous role of water in politics, accumulation processes and state-society-nature relations.

The right to water within the right to the city

In 2004–05, the 'World Charter for the Right to the City' (2005) was developed in Quito, Barcelona and Porto Alegre by networks associated with the World Social Forum. Its twelfth article had the following to say about water:

Right to Water and to Access and Supply of Domestic and Urban Public Services:

1. Cities should guarantee for all their citizens permanent access to public services of potable water, sanitation, waste removal, energy and telecommunications services, and facilities for health care, education, basic-goods supply, and recreation, in co-responsibility with other public or private bodies, in accordance with the legal framework established in international rights and by each country.
2. In regard to public services, cities should guarantee accessible social fees and adequate service for all persons including vulnerable persons or groups and the unemployed – even in the case of privatization of public services predating adoption of this Charter.
3. Cities should commit to guarantee that public services depend on the administrative level closest to the population, with citizen participation in their management and fiscal oversight. These services should remain under a legal regimen as public goods, impeding their privatization.
4. Cities should establish systems of social control over the quality of the services provided by public or private entities, in particular relative to quality control, cost determination, and attention to the public.

These suggest a layer of reforms with strong attention to technical and socially-just (if not necessarily ecological) considerations about water services, as well as subsidiarity and community control principles. But can the right to water be recast in more radical terms set out by urban revolutionaries such as Henri Lefebvre and David Harvey? The 'right to the city,' in Lefebvre's (1996) understanding is shaped by the class character of the project. Transforming questions of urban reform into ones shaped by an understanding of segregation, a proletarian conception of the 'right to the city' seeks to reconstruct an image of the whole within the created product or *oeuvre*. The fractured city is thus transformed into a whole that transforms the lived realities of those within it. Although not working on its own, the working class is central to the possibilities found within the right to the city for only 'this class, as a class, can decisively contribute to the reconstruction of centrality destroyed by a strategy of segregation found again in the menacing form of centres of decision-making' (Lefebvre, 1996, p154).

At a time in South Africa (and everywhere) when debate is intensifying about the alliances required to overthrow urban neoliberalism, as discussed below, we should heed Lefebvre's warning about the centrality of the working class to these struggles, but also Harvey's analysis of contradictions within accumulation. After all, in his *New Left Review* article on 'The Right to the City,' David Harvey (2008) draws upon the historical lessons of capital flows and urban form in mid-nineteenth century Paris and the post-war United States, before turning to the recent global property boom – which left very few cities untouched – and locating within it a profound and potentially unifying class struggle that also has important implications for water rights:

> A process of displacement and what I call 'accumulation by dispossession' lie at the core of urbanization under capitalism. It is the mirror-image of capital absorption through urban redevelopment, and is giving rise to numerous conflicts over the capture of valuable land from low-income populations that may have lived there for many years ... Since the urban process is a major channel of surplus use, establishing democratic management over its urban deployment constitutes the right to the city. Throughout capitalist history, some of the surplus value has been taxed, and in social-democratic phases the proportion at the state's disposal rose significantly. The neoliberal project over the last thirty years has been oriented towards privatizing that control ... One step towards unifying these struggles is to adopt the right to the city as both working slogan and political ideal, precisely because it focuses on the question of who commands the necessary connection between urbanization and surplus production and use. The democratization of that right, and the construction of a broad social movement to enforce its will is imperative

if the dispossessed are to take back the control which they have for so long been denied, and if they are to institute new modes of urbanization.

Contrast this analysis with a near-simultaneous statement – in a 2009 booklet, 'Systems of Cities: Integrating National and Local Policies, Connecting Institutions and Infrastructure' – from what many consider to be the brain of urban neoliberalism, the World Bank (2009). There is, to be sure, a confession that the neoliberal project was not successful in what the Bank had advertised since at least its 1986 New Urban Management policy (Bond, 2000). The document charts the rise of an 'enabling markets' approach to housing, based on policies developed within the World Bank based on the naïve assumption that this would somehow trickle down to low-income households. Nevertheless, by the mid-2000s even the Bank was prepared to admit that its approach had been 'far too sanguine about the difficulties' and was seeing a need for new policies and targeted subsidies for the urban poor. Nevertheless, within this 'Experience suggests that only a few regulations are critical: minimum plot sizes and minimum apartment sizes, limitations on floor area ratios, zoning plans that limit the type of use and the intensity of use of urban land, and land subdivision ratios of developable and saleable land in new greenfield developments' (World Bank, 2009).

Unlike Harvey, the World Bank has virtually nothing at all to say about 'rights' (except property rights and 'rights of way' for new roads and rail), and nothing at all to say about urban social movements. The closest is the document's reference to 'community-based organizations' which operate in 'partnerships' in Jamaica and Brazil to 'combine microfinance, land tenure, crime and violence prevention, investments in social infrastructure for day care, youth training, and health care with local community action and physical upgrading of slums.' Civil society in its most civilized form hence lubricates markets and acts as a social safety net for when municipal states fail.

Yet notwithstanding the confession, this discursive strategy leaves states with more scope to support markets, because rapid Third World urbanization generates market failures: *'The general principles of enabling markets are still valid, but must be combined with sensible policies and pragmatic approaches to urban planning and targeted subsidies for the urban poor.'* Recall that from the late 1980s, the World Bank had conclusively turned away from public housing and public services as central objectives of its lending and policy advice. Instead, the Bank drove its municipal partners to enhance the productivity of urban capital as it flowed through urban land markets (now enhanced by titles and registration), through housing finance systems (featuring solely private sector delivery and an end to state subsidies), through the much-celebrated (but extremely exploitative) informal economy, through (often newly-privatized) urban services such as transport, sewage, water and even primary health care services (via intensified cost-recovery), and the like. Recall, too, the rising barriers to access associated with the 1990s, turn to commercialized (sometimes privatized) urban water, electricity and transport services, and with the 2000s, real estate bubble. As a result, no matter the rhetoric now favouring 'targeted subsidies,' there are few cases where state financing has been sufficient to overcome the market-based barriers to the 'right to the city,' a point we will conclude with.

As Swyngedouw (2008, p3) pointed out, the context was a general realization about the limits to commodification in the private sector, if not the World Bank:

> This seems to be the world topsy-turvy. International and national governmental agencies insist on the market and the private sector as the main conduit to cure the world water's woes, while key private sector representatives retort that, despite great willingness to invest if the profit prospects are right, they cannot and will not take charge; the profits are just not forthcoming, the risks too high to manage, civil societies too demanding, contractual obligations too stringent, and subsidies have often been outlawed (the latter often exactly in order to produce a level playing field that permits open and fair competition).

These contradictions were especially important where social and natural processes overlapped. During the 1990s, the 'Integrated Water Resource Management' perspective began to focus on the nexus of bulk supply and retail water provision – in which water becomes an economic good first and foremost – but only to a very limited extent did it link consumption processes (especially overconsumption by firms and wealthy households) to ecosystem sustainability. Hence the rights of those affected by water extraction, especially those displaced by mega-dams that supplied cities like Johannesburg, have typically been ignored.

Making hydro-socio-ecological connections will be one of the crucial challenges for those invoking water rights. As Lefebvre (1996, p72) put it:

> Carried by the urban fabric, urban society and life penetrate the countryside. Such a way of living entails systems of objects and of values. The best known elements of the urban system of objects include water, electricity, gas (butane in the countryside), not to mention the car, the television, plastic utensils, 'modern' furniture, which entail new demands with regard to services.

Indeed, the ecological challenge of mobilizing water has, traditionally, been an important process of more general social and spatial organization (Strang, 2004). As Lefebvre (1996, p106) explained:

197

One knows that there was and there still is the oriental city, expression and projection on the ground, effect and cause, of the Asiatic mode of production; in this mode of production State power, resting on the city, organizes economically a more or less extensive agrarian zone, regulates and controls water, irrigation and drainage, the use of land, in brief, agricultural production.

Each different struggle for the right to the city is located within a specific political-economic context in which urbanization has been shaped by access to water. The early 'oriental despotism' that Karl Wittfogel (1957) discovered would follow from this Asiatic mode of production's emphasis on a strong central state's control of the water works gave way, in successive eras of city-building, to the central square role of water fountains in medieval market cities, and to huge infrastructural investments in capitalist cities. Within the latter, the neoliberal capitalist city has embarked on a variety of techniques that individualize and commodify water consumption, delinking it from its sources and disposal even though both these terrains are more difficult to accomplish through public-private partnerships. Given the emphasis on decentralization, as Bakker (2007, p436) suggests, 'The biophysical properties of resources, together with local governance frameworks, strongly influence the types of neoliberal reforms which are likely to be introduced.'

The next logical step on a civilizational ladder of water consumption would not, however, be simply a *Mazibuko*-style expansion of poor people's access (and technology) within the confines of the existing system. Acquiring a genuine right to water will require its 'commoning,' both horizontally across the populace, and vertically from the raindrop or borehole, all the way to the sewage outfall and the sea. But to get to the next mode of extraction, production, distribution, consumption and disposal of water requires a formidable social force to take us through and beyond rights, to the water commons.

The right to the city *and* to the water commons in South Africa

Tactically, anger about violations of water rights has taken forms ranging from direct protests, to informal/illegal reconnections and destruction of pre-payment meters, to a constitutional challenge over water services in Soweto. While having the potential to shift policy from market-based approaches to a narrative more conducive to 'social justice,' even in the face of powerful commercial interests and imperatives, the limits of a rights discourse are increasingly evident, as South Africa's 2008–09 courtroom dramas indicated. If the objective of those promoting the right to the city includes making water primarily an eco-social rather than a commercial good, these limits will have to be transcended. The need to encompass ecosystemic issues in rights discourses is illustrated by the enormous health impacts of unpurified water use (Global Health Watch, 2005, pp207–224).

Thus once we interrogate the limits to rights in the South African context, the most fruitful strategic approach may be to move from and beyond 'consumption-rights' to reinstate a notion of the commons, which includes broader hydropolitical systems. To do so, however, the South African struggle for water shows that social protests will need to intensify and ratchet up to force concessions that help remake the urban built environment. As expressed by David Harvey (2009), 'My argument is that if this crisis is basically a crisis of urbanization then the solution should be urbanization of a different sort and this is where the struggle for the right to the city becomes crucial because we have the opportunity to do something different.'

One of the first strategies, however, is defense. The struggle for water rights entails staying in place in the face of water disconnections and even evictions. Apartheid-era resistance to evictions is one precedent, but another is the moment in which a prior downturn in South Africa's 'Kuznets Cycle' (of roughly 15-year ups and downs in real estate prices) occurred, the early 1990s. The resulting 'negative equity' generated housing 'bonds boycotts' in South Africa's black townships. The few years of prior financial liberalization after 1985 combined with a class differentiation strategy by apartheid's rulers was manifest in the granting of 200,000 mortgage bonds to first-time black borrowers over the subsequent four years. But the long 1989–93 recession left 500,000 freshly unemployed workers and their families unable to pay for housing. This in turn helped generate a collective refusal to repay housing bonds until certain conditions were met. The tactic moved from the site of the Uitenhage Volkswagen auto strike in the Eastern Cape to the Johannesburg area in 1990, as a consequence of two factors: shoddy housing construction (for which the homebuyers had no other means of recourse than boycotting the housing bond) and the rise in interest rates from 12.5 per cent (−6 per cent in real terms) in 1988 to 21 per cent (+7 per cent in real terms) in late 1989, which in most cases doubled monthly bond repayments (Bond, 2000).

As a result of the resistance, township housing foreclosures which could not be consummated due to refusal of the defaulting borrowers (supported by the community) to vacate their houses, and the leading financier's US$700 million black housing bond exposure in September 1992 was the reason that its holding company (Nedcor) lost 20 per cent of its Johannesburg Stock Exchange share value (in excess of US$150 million lost) in a single week, following a threat of a national bond boycott from the national civic organization. Locally, if a bank did bring in a sheriff to foreclose and evict defaulters, it was not uncommon for a street committee of activists to burn the house down before the new owners completed the purchase and moved in. Such power, in turn, allowed both the national and local civic associations to negotiate concessions from the banks (Mayekiso, 1996).

However, there are few links between the early 1990s civics which used these micro-Polanyian tactics successfully, and the 2000s generation of 'new

199

social movements' which shifted to decommodification of water and electricity through illegal reconnections (Desai, 2002). The differences partly reflect how little of the late 2000s, mobilizing opportunities came from formal sector housing, and instead related to higher utility bills or forced removals of shack settlements. Still, there are profound lessons from the recent upsurge of social activism for resistance not only to the implications of world capitalist crisis in South Africa, but elsewhere.

The lessons come from deglobalization and decommodification strategies used to acquire basic needs goods, as exemplified in South Africa by the national Treatment Action Campaign (TAC) and Johannesburg Anti-Privatization Forum which have won, respectively, antiretroviral medicines needed to fight AIDS and publicly-provided water (Bond, 2006). The drugs are now made locally in Africa – in Johannesburg, Kampala, Harare, and so on – and on a generic not a branded basis, and generally provided free of charge, a great advance upon the US$15,000/patient/year cost of branded AIDS medicines a decade earlier (in South Africa, half a million people receive them). The right to healthcare in the South African city, hence, requires the communing of intellectual property rights, which were successfully achieved by the TAC by mid-decade in the 2000s after extreme resistance was required.

The ability of social movements such as in the health, water and housing sectors to win major concessions from the capitalist state's courts under conditions of crisis is hotly contested, and will have further implications for movement strategies in the months ahead. Marie Huchzermeyer (2009, pp3–4) argues that the Constitution mandates 'an equal right to the city.' However:

It was only in 2000 that the Bill of Rights was evoked by a marginalized and violated urban community (represented by Irene Grootboom) in the Constitutional Court. In what was received as a landmark ruling, the Court interfered with the Executive, instructing the Ministry of Housing to amend its housing policy to better cater for those living in intolerable conditions. It took 4 further years for the policy changes to be adopted into housing policy. Chapters 12 and 13 were added to the national Housing Code: Housing in Emergency Circumstances and Upgrading of Informal Settlements. In the following 5 years, these two policies have not been properly implemented, if at all. Unnecessary violations have continued and marginalized communities have had to resort to the courts. However, the landscape has changed significantly. Whereas the *Grootboom* case involved an isolated community with only a loose network of support through the Legal Resources Centre which acted as 'Friends of the Court', today cases reach the Constitutional Court through social movements such as Landless People's Movement, Inner City Tenant Forum, Abahlali base Mjondolo, Anti-Privatization Forum and the Anti-Eviction Campaign. These movements coordinate, exchange, and take an interest in one another's legal struggles.

Huchzermeyer (2009, p4) suggests this strategy fills a 'gap in left thinking about the city (the gap derived from the Marxist ideology of nothing but a revolution)' and that the 'Right to the City' movement articulated by Lefebvre and Harvey should include marginal gains through courts: 'Urban Reform in this sense is a pragmatic commitment to gradual but radical change towards grassroots autonomy as a basis for equal rights.' After all, 'three components of the right to the city – equal participation in decision-making, equal access to and use of the city and equal access to basic services – have all been brought before the Constitutional Court through a coalition between grassroots social movements and a sympathetic middle class network' (even though 'this language is fast being usurped by the mainstream within the UN, UN-Habitat, NGOs, think tanks, consultants etc., in something of an empty buzz word, where the concept of grassroots autonomy and meaningful convergence is completely forgotten').

As we have seen, however, critics point to the opposite processes in the water case, and consider a move through and beyond human rights rhetoric necessary on grounds not only that – following the Critical Legal Scholarship tradition – rights talk is only conjuncturally and contingently useful (Roithmayr, 2011). Ashwin Desai (2010) offers some powerful considerations about the danger of legalism when building the South African urban social movements:

> If one surveys the jurisprudence of how socio-economic rights have been approached by our courts there is, despite all the chatter, one central and striking feature. Cases where the decision would have caused government substantial outlay of money or a major change in how they make their gross budgetary allocations, have all been lost. Cases where money was not the issue such as the *TAC* case or where what was being asked for was essentially negative – to be left alone – the courts have at times come grandly to the aid of the poor. And even to get some of these judgments enforced by the executive is a story in and of itself. I have no problems using the law defensively but when it comes to constitute the norms by which political advances are determined, it is extremely dangerous. By flirting with legalism, movements have had their demands become infected with court pleadings. We have heartfelt pleas for the observance of purely procedural stuff, consult us before you evict us. We have demands for housing, now become 'in situ upgrading' and 'reasonable' government action.

In addition, the limits of neoliberal capitalist democracy sometimes stand exposed, when battles between grassroots-based social movements and the state must be decided in a manner cognizant of the costs of labor power's reproduction. At that point, if a demand upon the state to provide much greater subsidies to working-class people in turn impinges upon capital's (and rich people's) prerogatives, we can expect rejection, in much the same way Rod Burgess (1978) criticized an earlier version of relatively unambitious Urban Reform (John Turner's self-help housing), on grounds that it fit into – not fought against – the process by which capital lowered its labor reproduction costs. It may be too early to tell whether court victories won by social movements for AIDS medicines and housing access are the more durable pattern that reifies rights talk, or whether the defeat of the Soweto water-rights movement is more typical. Sceptics of rights talk suggest, instead, a 'Commons' strategy, by way of resource sharing and illegal commandeering of water pipes and electricity lines during times of crisis (Bakker's chapter in this book; Bond, 2002; Desai, 2002; Naidoo, 2009; Ngwane, 2009).

The challenge for South Africans committed to a different society, economy and city is combining requisite humility based upon the limited gains social movements have won so far (in many cases matched by the worsening of regular defeats) with the soaring ambitions required to match the scale of the systemic crisis and the extent of social protest. Looking retrospectively, it is easy to see that the independent left – radical urban social movements, the landless movement, serious environmentalists and the left intelligentsia – peaked too early, in the impressive marches against Durban's World Conference Against Racism in 2001 and Johannesburg's World Summit on Sustainable Development in 2002. The 2003 protests against the US/UK for the Iraq war were impressive, too. But in retrospect, although in each case they out-organized the Alliance, the harsh reality of weak local organization outside the three largest cities – plus interminable splits within the community, labor and environmental left – allowed for a steady decline in subsequent years.

The irony is that the upsurge of recent protest of a 'popcorn' character – i.e., rising quickly in all directions but then immediately subsiding – screams out for the kind of organization that once worked so well in parts of Johannesburg, Durban and Cape Town. The radical urban movements have not jumped in to effectively marshal or even join thousands of 'service delivery protests' and trade union strikes and student revolts and environmental critiques of the past years. The independent left's organizers and intelligentsia have so far been unable to inject a structural analysis into the protest narratives, or to help network this discontent.

Moreover, there are ideological, strategic and material problems that South Africa's independent left has failed to overcome, including the division between autonomist and socialist currents, and the lack of mutual respect for various left traditions, including Trotskyism, anarchism, Black Consciousness and feminism. A synthetic approach still appears impossible in the contemporary moment. Aside from a campaign against a $3.75 billion World Bank loan to Eskom that unites red (including labor and community) and green against electricity privatization, extreme price increases (127 per cent in real terms over four years) and climate damage, nor do strategic convergences appear obvious. For example, one strategic problem – capable of dividing major urban social movements – is whether to field candidates at elections. Another problem is the independent left's reliance upon a few radical funding sources instead of following trade union traditions by raising funds from members (the willingness of German voters to vote Die Linke may have more than a little influence on the South African left).

By all accounts, the crucial leap forward will be when leftist trade unions and the more serious South Africa's Communist Party members ally with the independent left. The big question is, when will Cosatu reach the limits of their project within the Alliance. Many had anticipated the showdown in 2007 to go badly for unionists and communists, and they (myself included) were proven very wrong. There is probably no better national trade union movement in the English-speaking world than Cosatu, so that error requires a rapid correction. By March 2010, after a disappointing State of the Nation speech by Zuma followed by a reactionary budget speech that opened up a two-tier labor market (characterized by hated labor-broking outsourcing) and retained orthodox monetary policy, the showdown appears much closer. It may hinge around Zuma's alliance with his radical-sounding youth, led by Julius Malema, whose 'tenderpreneur' skills in accessing state contracts reeked of corruption.

These challenges are not particularly new nor unique, with many leftists in Latin America and Asia reporting similar opportunities during this crisis but profound barriers to making the decisive gains anticipated. It is, however, in South Africa's intense confrontations during capitalist crisis that we may soon see, as we did in the mid-1980s and early 2000s, a resurgence of perhaps the world's most impressive urban social movements. And if not, we may see a degeneration into far worse conditions than even now prevail, in a post-apartheid South Africa more economically unequal, more environmentally unsustainable and more justified in fostering anger-ridden grassroots expectations, than during apartheid itself.

References

Bakker, K. (2007) 'The "commons" versus the "commodity": alter-globalization, anti-privatization and the human right to water in the global South', *Antipode*, vol 39, no 3, pp430–455

Bond, P. (2000) *Cities of Gold, Townships of Coal*, Africa World Press, Trenton

Bond, P. (2002) *Unsustainable South Africa*, Merlin Press, London

Bond, P. (2006) *Talk Left Walk Right*, University of KwaZulu-Natal Press, Pietermaritzburg

Bond, P. (2009) 'Repaying Africa for climate crisis: ecological debt as development finance alternative to carbon trading', in S. Böhm and S. Dabhi (eds) *Upsetting the Offset: The Political Economy of Carbon Markets*, MayFlyBooks, London

Bond, P. (2010) 'Fighting for the right to the city: discursive and political lessons from the right to water', Plenary Lecture at Right to Water Conference, Syracuse University, New York, 29–30 March 2010

Bond, P. and J. Dugard (2008) 'The case of Johannesburg water: what really happened at the pre-paid "parish pump"', *Law, Democracy and Development*, vol 12, no 1, pp1–28

Brand, D. (2005) 'The politics of need interpretation and the adjudication of socio-economic rights claims in South Africa', in A.J. van der Walt (ed) *Theories of Social and Economic Justice*, Stellenbosch University Press, Stellenbosch

Burgess, R. (1978) 'Petty commodity housing or dweller control?', *World Development*, vol 6, no 9/10, pp1105–1133

Centre for Applied Legal Studies (2009) 'Press Statement', Johannesburg, 25 March

Coalition Against Water Privatization (2009) 'Press Statement: "Phiri Water Case: Constitutional Court Fails the Poor and the Constitution"', Johannesburg, 2 October

Danchin, P. (2010) 'A human right to water? The South African Constitutional Court's decision in the Mazibuko Case', *EJIL Talk, European Journal of International Law*, 13 January, www.ejiltalk.org/a-human-right-to-water-the-south-african-constitutional-court's-decision-in-the-mazibuko-case

Desai, A. (2002) *We Are the Poors*, Monthly Review Press, New York

Desai, A. (2010) 'The state of the social movements', Presented to the CCS/Wolpe Lecture Panel 'Social justice ideas in civil society politics, global and local: A colloquium of scholar-activists', Centre for Civil Society, Durban, 29 July 2010

Dugard, J. (2010a) 'Civic action and legal mobilisation: the Phiri Water Meters Case', in J. Handmaker and R. Berkhout (eds), *Mobilising Social Justice in South Africa: Perspectives from Researchers and Practitioners*, Pretoria University Law Press, Pretoria

Dugard, J. (2010b) 'Reply', *EJIL Talk, European Journal of International Law*, 17 April, www.ejiltalk.org/a-human-right-to-water-the-south-african-constitutional-court's-decision-in-the-mazibuko-case

Global Health Watch (2005) *Global Health Watch*, Zed Books, London

Harvey, D. (2008) 'The right to the city', *New Left Review*, no 53

Harvey, D. (2009) Opening speech at the Urban Reform Tent, World Social Forum, Belem, 29 January

Huchzermeyer, M. (2009) 'Does recent litigation bring us any closer to a right to the city?', Paper presented at the University of Johannesburg workshop on 'Intellectuals, ideology, protests and civil society', 30 October 2009

Lefebvre, H. (1996) *Writings on Cities*, Basil Blackwell, Oxford

Madlingozi, T. (2007) 'Good victim, bad victim: apartheid's beneficiaries, victims and the struggle for social justice', in W. le Roux and K. van Marle (eds) *Law, Memory and the Legacy of Apartheid: Ten Years after AZAPO v President of South Africa*, University of Pretoria Press, Pretoria

Mayekiso, M. (1996) *Townships Politics*, Monthly Review, New York

Mazibuko & Others v the City of Johannesburg & Others (2008) Unreported case no 06/13865 in the Johannesburg High Court

Naidoo, P. (2009) 'The making of "The Poor" in post-apartheid South Africa', Masters research thesis, University of KwaZulu-Natal School of Development Studies, Durban

Ngwane, T. (2009) 'Ideology and agency in protest politics', Masters research thesis proposal, University of KwaZulu-Natal School of Development Studies, Durban

Pieterse, M. (2007) 'Eating socioeconomic rights: the usefulness of rights talk in alleviating social hardship revisited', *Human Rights Quarterly*, vol 29, pp796–822

Republic of South Africa (1996) *Constitution of the Republic of South Africa Act 108 of 1996*, Cape Town

Roithmayr, D. (2011) 'Lessons from *Mazibuko*: persistent inequality and the commons', *Constitutional Court Review*, 1

Strang, V. (2004) *The Meaning of Water*, Berg Publishers, Oxford

Swyngedouw, E. (2008) 'Retooling the Washington consensus: the contradictions of H_2O under neo-liberalism and the tyranny of participatory governance', Paper presented to the Centre for Civil Society, Durban, 3 July 2008

United Nations Educational, Social and Cultural Organization (2006) *World Water Development Report 2: Water a Shared Responsibility*, Berghahn Books, New York and Oxford www.unesco.org/water/wwap/wwdr/wwdr2/

Wittogel, K. (1957) *Oriental Despotism: A Comparative Study of Total Power*, Yale University Press, New Haven

World Bank (2009) 'Systems of Cities: Integrating National and Local Policies, Connecting Institutions and Infrastructure', Washington, DC, Sustainable Development Network

World Charter for the Right to the City (2005) Porto Alegre, www.urbanreinventors. net/3/wsf.pdf

13

ANTI-PRIVATIZATION STRUGGLES AND THE RIGHT TO WATER IN INDIA

Engendering cultures of opposition

Krista Bywater

They came to our village with glittering offers; that our people would get ample job opportunities in the plant; the overall development of our village would be taken care of. . . . On the contrary, six months went by, slowly we started facing the reverse effects. Our precious water resource had been stolen. . . . Where would I get some fresh and pure drinking water anymore? How many kilometers should we have to walk to fetch a drop of water? Who will compensate the heavy loss incurred upon us by this giant cola plant?

> (*Mylamma, member of the Anti-Coca-Cola People's Struggle Committee, Plachimada, Kerala*)[1]

Water is something that nobody can live without. So then what's going to happen when people can't get water? People will be fighting for their lives. They will have no choice, especially the poorest people who have nothing to lose. How can they live? So they will go out and risk their lives for this. There can easily be violence in the streets, and no one in Delhi wants this . . .

> (*Beena, member of the Right to Water Campaign, Delhi, India*)

Introduction

Thousands of protesters campaigned in Plachimada, Kerala, and Delhi, India, to oppose the private control of public water supplies and to ensure the right to water. Activists in Plachimada successfully shut down a Coca-Cola[2] plant that depleted the village's groundwater, and the popular movement in Delhi stopped the privatization of the city's water utility. The struggles in Plachimada and Delhi are not uncommon as rural communities all over the world are increasingly competing with industries for access to local water supplies, and urban residents are defending water utilities against private companies that seek new customers and markets. What distinguishes these cases from other water conflicts is that the broad-based movements generated enough political pressure to force the Indian state and central governments to ensure that water would be managed in the public's interest and not for profit.

People around the world look to Plachimada as an example of a success-ful people's movement in which Adivasis,[3] a population with little power and few resources, forced Coke, a huge transnational corporation, to shut down. Likewise, activists characterize the struggle in Delhi as a successful anti-privatization movement that prevented a World Bank-sponsored project. How can we account for the rise of these water conflicts and movements? And what can these campaigns teach us about establishing the right to water? Using ethnographic data, this chapter demonstrates that the environ-mental movements gained popular support and successfully stopped the water privatization schemes because they: promoted understandings of water as a common resource and human right, drew upon idioms such as 'water for life, not for profit', and publicized anti-globalization discourses and negative international experiences of water privatization. The two strug-gles also combined global discourses with grassroots organizing to address the cultural, economic, and political realities of the affected communities. The movements in Plachimada and Delhi prove that integrating human-rights discourse, conceptions of water commons, and cultural understandings of water into campaigns can advance claims to the right to water.

The cases in Plachimada and Delhi challenge Bakker's position (2007 and in this book) which emphasizes the limitations of the human rights framework to counter water privatization. Bakker advocates for abandoning human rights discourse and calls on activists to use notions of water commons to prevent water privatization and to ensure that governments and communities meet people's water needs. For Bakker only positing that resources are public and communal directly challenges the regime of private rights and private control of resources. Yet, analysis of the two cases in India shows how a multidimensional approach that includes promoting water as a human right can advance water claims. Additionally, this chapter offers lessons about the importance of culture in environmental organizing and how people engaged in similar conflicts can establish the right to water.[4]

The chapter begins with a description of the struggles in Plachimada and Delhi to contextualize the water conflicts. Then I provide the theoretical background and highlight the importance of considering culture and politics when analyzing environmental movements. I offer Foran's concept of polit-ical cultures of opposition (PCOs) as a valuable tool to distill the central elements of successful claims to the right to water. In the remaining sections,

I evaluate the water movements and explain how each struggle stopped the water privatization projects. The conclusion underscores that a multipronged approach, which promotes: the multiple meanings of water (including water as a human right and water as commons), broad networks, and global connections can advance claims to the right to water.

Water conflicts in Plachimada and Delhi

On Earth Day in 2002, approximately 1,500 protesters formed a human chain in front of Coca-Cola's bottling plant in the small hamlet of Plachimada, Kerala. Adivasi men and women organized the *Coca-Cola Virudha Janakeeya Samara Samithi* (Anti-Coca-Cola People's Struggle Committee) and demanded that 'Coca-Cola Quit Plachimada, Quit India'. In a state known for its heavy monsoons, protesters asserted that the company's practices depleted and polluted the community's land and only source of water – the ground water. Protesters including Mylamma also charged Coke with selling its toxic waste – a sludge – to unsuspecting farmers as a fertilizer. Over the next two years people and organizations in Kerala and abroad began to support the movement, and it grew and strengthened. As the struggle escalated it developed additional dimensions; anti-Coke protesters charged the company with privatizing the community's ground water and destroying people's livelihoods. In the process the struggle became not only a fight for the community's survival but an environmental, anti-globalization, and anti-privatization movement.

Due to public opposition the Coca-Cola plant stopped production in March 2004, but the anti-Coke struggle continues today. On March 22, 2010, World Water Day, a high level committee formed by the Kerala state government found Coke liable for $48 million for the damage its plant caused in Plachimada. Although the movement has obtained some success, residents will likely wait a long time for any reparations, and those living near the factory still suffer from water shortages and pollution. The case offers a complicated story that involves degrees of success and failure, numerous legal battles, and multiple social actors. Nine years after the conflict started protesters still occupy the movement's *samarapanthal* or protest shed across from the Coke plant and demand that the company be held criminally liable for its actions.

In 2004, the same year that villagers in Plachimada forced the Coca-Cola plant to close, residents in Delhi began their own anti-privatization movement. Thousands of people including students, women's groups, left political parties, trade unions, non-governmental organizations (NGOs), resident welfare associations, and citizens from all segments of society protested Delhi's Water Supply and Sewage Project. Groups such as college students and slum residents, who rarely work together, joined forces to resist the water scheme commonly known as Delhi's water privatization project. The protesters united to prevent the Delhi *Jal* Board (Delhi Water Board) from pursuing a $140 million World Bank loan to restructure the water utility and allow foreign private firms to manage the city's water and sanitation services. Protesters stated that allowing private companies to manage the utility for profit, eliminating free water supplies, and increasing user costs constituted privatization and that these reforms would not improve services and would unfairly harm the city's poor residents. Together dissenters formed Delhi's anti-privatization water movement and demanded that the Indian government remain in control of the city's water sector. The opposition in Delhi was simultaneously an anti-privatization movement that fought to keep the water utility public and an environmental movement that advocated for sustainable management of the city's water resources.

Most protesters and organizations worked collectively under two campaigns: the Right to Water Campaign and the Citizens Front for Water Democracy. Groups with each of these coalitions joined forces and held dozens of protests, *dharnas* (sit-ins in front of political offices), and marches. They also toiled individually to educate Delhi's residents about the water development plan and to build opposition against the project. India's national newspapers published articles such as *The Hindu*'s piece 'A Front against Water Privatization', which detailed the opposition to the project, and left political parties including the Communist Party of India (Marxists) wrote open letters to Prime Minister Manmohan Singh denouncing the project (Karat et al, 2005). As hundreds of people filled the streets during protests throughout 2004 and 2005 and as media reported on the water privatization scheme, the city became embroiled in one of the country's biggest anti-privatization water movements. The Delhi *Jal* Board (DJB) ultimately withdrew its loan request from the World Bank because of the public opposition to the project.

Both of the struggles successfully prevented what protesters understood to be the privatization of public water supplies.[5] Each of the movements achieved their initial objectives; in Kerala the ground water is again communally governed, and in Delhi the water utility remains government-run. These are compelling case studies because, despite powerful resistance, campaigners pressured government officials to protect water resources from private control and to act on behalf of the public.

As noted at the beginning of this chapter, the accomplishments of the two anti-privatization movements in India raise several questions. What can we learn from these conflicts about achieving the right to water? And how did people with few resources stop a huge transnational corporation and a World Bank-sponsored development project? In order to answer these questions it is first necessary to highlight critical aspects of environmental movements – culture and politics.

Theoretical underpinnings: culture, politics, and coalitions

The growing literature on water problems emphasizes that attempts to understand disputes must examine the impact of economic systems, geographies, historical formations, power relations, social inequalities, and everyday water experiences (Swyngedouw, 1995, 2004; Crow and Sultana, 2002; Bakker, 2003, 2007; Coles and Wallace, 2005; Gandy, 2008; Sultana, 2011; Truelove, 2011) as well as the cultural meanings and values of water (Shiva, 2002; Strang, 2006; Baviskar, 2007; Iyer, 2007). The increasingly interdisciplinary nature of recent scholarship reveals the complexity of water conflicts and numerous possibilities for achieving the right to water.

Within the plurality of ways of examining water struggles, the role of culture in water conflicts is an area that deserves further attention (Baviskar, 2007). As a result of scholarly analyses, there is growing awareness among development practitioners that cultural politics can help to explain why resource conflicts and environmental movements take place. For instance, people often attribute different meanings to resources and these meanings can clash with dominant cultural understandings (such as those reflected in development schemes) and struggles can ensue (for examples see Agrawal, 2005; Strang, 2006; Baviskar, 2007). Cultural politics (Alvarez et al, 1998), critical cultural perspectives (Swyngedouw, 1999), and cultural opposition (Strang, 2006) are related ideas used to explain the rise of resource movements. Although scholars continue to analyze the role of culture in water conflicts, further investigation can: improve understandings of the social conditions that foster water disputes, help explain why particular cultural oppositions achieve movement goals, and elucidate why others fail to establish the right to water.

A concept that can enhance analyses of resource conflicts is John Foran's notion of political cultures of opposition (PCOs). He states that strong oppositional political cultures helps to explain 'the emergence of radical new struggles in the twenty-first century' including the global justice movement (Foran, 2009, p.144) and global anti-privatization movement. While Foran developed the concept to account for successful social revolutions, PCOs helps to elucidate how political cultures can engender successful water struggles and claims to the right to water. Within his framework culture is understood to be political and oppositional when meanings are constitutive of social processes that challenge the status quo and seek to redress unequal power relations. Foran and Reed (2002, p.339) build upon the scholarship of critical theorists like Stuart Hall and Raymond Williams and observe that the concept is similar to the ideas of:

> Charles Tilly, with his passing but suggestive reference to 'cultural repertoires of revolution' (1978:151–9,224–5), Ann Swidler, with her influential metaphor of a 'cultural tool-kit' (1986), James Scott's

work on 'hidden transcripts' (1990), or even C. Wright Mills's 'vocabularies of motives' (1963[1940]:442).

Foran's formulation of PCOs is valuable because it reveals the features of popular cultural opposition that are at times obfuscated in public discussions of environmental movements. There are several dimensions of political cultures of opposition, which distinguish these formations and enable them to produce strong progressive movements. First, PCOs originate in shared 'structures of feeling' (Williams, 1960), experiences, ideologies, cultural idioms, organizations, and networks. They '. . . are forged out of the encounter of these different elements and can become generalized when groups of people organize themselves into networks and organizations seeking to change the established order' (Foran, 2009, p.146). Second, Foran notes that several PCOs can exist within any society or social movement, and movements often succeed when they can synthesize various PCOs into an effective slogan, goal, or overarching perspective such as 'reduce, reuse, recycle'.

Third, PCOs can help to unite a broad coalition of actors. People across social groups and from different classes, religions, ethnicities/castes, and genders can bond through a shared political culture of opposition. Collective understandings have the power to bridge social differences, which facilitates the formation of diverse and far-reaching alliances such as the blue green coalitions within the anti-globalization movement. Finally, Foran states that political cultures of opposition are emerging within the contemporary global context. PCOs focus on emotions such as love, experiences of globalization from above, cultural idioms (with an emphasis on deeply democratic strategies and goals, inclusiveness, non-violence, dignity, social justice, etc.), and new forms of organizations (broad coalitions and horizontal networks) (2009, p.146). Movements for radical social change are in transition, as the twentieth-century revolutions waged by socialist political parties, guerrilla armies, and religious leaderships are being replaced by direct action groups, horizontal networks, and new types of left-green political parties. If political cultures of opposition become popular, they offer a powerful way to spur wide-spread political participation and pressure governments and officials to work in the interest of the public.

Applying Foran's concept of PCOs to environmental movements helps to evaluate the ways people make sense of and challenge dominant environmental discourses, knowledge, and perceptions. Understandings of environmental resources that seek to subvert and alter hegemonic views of nature and society-environment relations can cultivate opportunities for alternative forms of resource governance. The following sections utilize PCOs to analyze the emergence and unlikely success of the movements in Plachimada and Delhi.

Culture and campaigns: the meanings of water

The central goal of the two Indian movements was to stop the water privatization projects in order to protect public and communally governed water supplies. To achieve this objective, activists reasoned that they would need broad-based support to pressure political officials to meet their demands. This section examines how the struggles appealed to cultural understandings of water in India and encouraged people to join the anti-privatization and environmental movements. Political cultures of opposition developed as the resistance in Plachimada and Delhi drew on perceptions of water as a cultural and common resource and human right. In both cases, protesters found treating water like a commodity and the private control of resources to be contrary to their cultural understandings of the resource.

During interviews and countless conversations with people in and outside of the water movements, common interpretations of water emerged. Many of the activists explained that they joined the struggles because they believed that the World Bank, DJB, and Coca-Cola Company ignored the cultural (water is a common resource), religious (water is sacred), moral (water is life sustaining), and legal (water is a human right and water provision is a government's responsibility) meanings of water. Beena, a trade union activist and member of the Citizen's Front for Water Democracy, explained how the campaign utilized the discourse of water as a human right. Her comments represent how people typically discussed the importance of a rights-based approach:

> Of course, we said water is life for us. . . . [W]ater is life for us, and it's very, very important for us. . . . So it should be handled appropriately. . . . No water privatization should take place. Water should be the responsibility of governments and it should be the right of people. And the government has to fulfill their obligation to supply water to the people.

Beena emphasizes that water is a human right and that it is the legal 'obligation' of governments to provide water to their citizens. By stressing that 'water is life', she also suggests that water access cannot be separated from the right to life. To deny someone water then raises moral dilemmas which challenge the idea that water can be commodified and treated like any other good. These understandings of water are reflected in campaign materials produced by the Citizen's Front for Water Democracy including the *Declaration of the People's World Water Movement*, which states 'water is life', 'water is sacred', and is a human right.

Neelam, another activist and organizer in Delhi's anti-privatization movement, identified the significance of the cultural meanings of water. Her comments mirror the refrains told to me by the majority of protesters I encountered in Delhi. In the selection below, she explains how the conflict over the meanings of water facilitated the popular resistance:

> And part of the resistance I think is also anchored in this concept of water whether it's a commodity or whether it's a right. Because it's part of our Delhi culture, not just livelihood. Of course it's [about] livelihood, but [it's also about] the place of water in our cultures and how we look at it. And that's where a lot of the language of the resistance is anchored in that particular concept. . . . The language that water cannot be commodified.

In the quote Neelam recognizes that water is a human right that ensures survival, but she also points to the need to recognize the 'place of water in our cultures and how we look at it'. People view water as culturally important and imbue it with social significance. The cultural meanings of water are incompatible with privatization schemes that prioritize water as a commodity. People's human and cultural rights to water and government's moral obligation to ensure people's livelihoods informed the campaigns' language of resistance. These ideas resonated with the public and persuaded people from a cross section of society to join the opposition.

It is sometimes difficult for people to appreciate the cultural politics of water and its multiple meanings. A myriad of scholars including Crow, Lindquist, and Wilson (1995), Shiva (2002), Iyer (2007), and Baviskar (2007) have documented the sacredness of water and water resources as well as the religious rituals and uses of water in India. Of course not all people in India believe water is a sacred resource, but it is a cultural norm to appreciate the multiple meanings people attribute to water. Many protesters like Neelam joined the anti-privatization movements because they thought the Delhi project and Coca-Cola Company treated water like an economic good, which they found to be incompatible with the religious, moral, cultural and legal meanings of the resource. Commenting on the different perspectives of water Iyer writes:

> Water is perceived by different people (or by the same people in different contexts) in different ways: as a *commodity*, a *commons*, as a *basic right* and a sacred resource or *divinity*. . . . those who regard water as 'commons' or a 'common pool resource' tend to deny vehemently that it is a commodity. Contrariwise, those who see water as a commodity are often blind to the other dimensions of water.

> (2007, p.77)

The Coca-Cola Company, World Bank, and Delhi *Jal* Board were blind to the cultural significance and multiple meanings of water. The water conflicts erupted as people organized and articulated political cultures of opposition that contested the development schemes which sought to commodify and privatize water.

The anti-privatization movements upheld the multiple meanings of water and their campaigning resonated with Delhi's residents and citizens throughout Kerala. Idioms such as 'water for life, not for profit' and 'water is our birth right' popularized the anti-privatization movements and represented the PCOs that formed to contest the water privatization in Plachimada and Delhi. From the common refrains used by protesters, it is evident that society-environment relations or how people manage the environment is affected by the meanings people attribute to resources. One of the most important idioms used in the movements was the declaration that 'water is a human right'. As mentioned earlier, Bakker (2007 and in this book) maintains that human rights and privatization are not incompatible. Indeed corporations in the water business, like Suez and Vivendi, and rightwing think tanks assert that water is a human right that can be met through privatization (Bakker, 2007, p.440). Therefore, she suggests that anti-privatization activists replace their emphasis on universal rights with a focus on water commons. While human rights are not a theoretical counterpoint to privatization, the case studies demonstrate that rights discourse is useful because it invokes political cultures of opposition that rally public support. Few people deny that water is a human right, and the rights discourse appealed to the general public and helped protesters generate support.

Another particularly powerful slogan employed by protesters in the Anti-Coke struggle was 'Coca-Cola Quit Plachimada, Quit India'. It drew on a political culture of opposition known throughout India and much of the world – Gandhi's independence movement. The idiom refers to Gandhi's Quit India Movement and his fight against British colonialism and imperialism. The slogan succinctly captures the protesters' opposition to the Coca-Cola Company. Campaigners likened the company to a foreign power that was unjustly usurping and destroying Indian resources for its own profit. The movement deemed Coke's operations in Plachimada as a form of foreign imperialism that was exploiting the village's water resources. Getting Coke out of Plachimada became synonymous with India's decolonization and fight against the British. The slogan endorsed Indian nationalism and protests against the foreign control of resources. This idiom provided a particularly useful analogy to motivate people to join the movement. It associated the anti-Coke struggle with the noble and revered Gandhi and the ethos of the Quit India Movement. The simple phrase 'Coca-Cola Quit Plachimada, Quit India' decried the foreign control of resources and demanded that Coke leave the country. The campaigns in Plachimada and Delhi synthesized cultural understandings of water and the ethos of Gandhi's Quit India Movement into catchy campaign slogans. These idioms helped to generate political cultures of opposition that attracted a cross-section of society to support the movements.

Broad coalitions and popular movements

Thus far I have analyzed how the movements used people's multiple under-standings of water to garner popular support for the anti-privatization struggles. This section details how protesters drew on people's experiences and tapped into existing political cultures of opposition to motivate the public to join the movements. The Plachimada struggle capitalized on the culture of activism and left politics in the state of Kerala and attracted a diverse array of supporters. I asked scholars and water activists around India why the anti-Coke struggle in Kerala was able to shut down the company's operations while other anti-Coke movements, like the ones in Mehndiganj and Jaipur, had not attained the same success. People always stressed that Kerala's unique development including its history of left politics and strong *panchayat* (village council) system, encouraged political participa-tion unparalleled in the rest of India. This culture of resistance and political participation made it easy for the anti-Coke movement to gain supporters. In a conversation with Abhay, a former Coca-Cola employee, he explained:

> It was hard to keep working with the movement going on. So many people here went against the company. Here we have a strong *panchayat*. . . . When it also went against Coke the struggle became even more popular. Because of the protests the *panchayat* took away the company's business license.

Abhay went on to say that he did not join the movement and continued to work at the factory in order to financially support his family. As he states the *panchayat* did cancel Coke's business license because of the opposition. Kerala's Supreme Court later overturned the council's decision, but the movement gained momentum from the political support of the council. Kerala has the strongest *panchayat* system in India since the state's political system is decentralized with governance responsibilities and finances equally shared by local councils and the state government. The *panchayat* system encourages direct political participation at the village level and councils are charged with representing local residents. Protesters in Plachimada petitioned council members to represent the needs of residents and eventually created enough political pressure to force Coca-Cola's bottling plant to close.

Scholarship on Kerala reinforces interviewees' claims of the unique culture of democratic participation and activism in the state. Franke and Chasin (1994, 2000) and Parayil (2000) are among the numerous scholars who high-light Kerala's unique development model, which is marked by: 1) a series of people's movements and government reforms that redistributed wealth and resources, 2) exceptionally high quality of life standards despite low economic growth, and 3) a culture of activism and democratic political participation in the general populace and among leaders. Kerala's alternative model of development has resulted in a highly literate, educated, politically left, and socially engaged populace. This long history of social activism in Kerala encouraged people in Plachimada to take up the daunting task of protesting against the giant Coca-Cola Corporation.

Moreover, many activist networks were established throughout the state prior to Plachimada's conflict, and organizations simply added the anti-Coke struggle to their agendas. Evidence of broad-based support for the movement is found in the number of organizations that aided Adivasis in their struggle. Dozens of groups including the Solidarity Youth Movement, People's Union for Civil Liberty (PUCL), and *Harita* Development Association joined the anti-Coke movement. One of the most important horizontal networks that formed was the *Plachimada Aikyadardhya Samithi* (Plachimada Solidarity Committee). More than thirty organizations have joined the committee, donated money, given legal advice, and participated in anti-Coke rallies in Plachimada and Trivandrum – the state capital. The primary mission of the committee is to support Adivasis and other locals affected by Coke's operations. The anti-Coke struggle welcomed the outside assistance and has worked closely with the organizations to stop the privatization of the community's water. Adivasis gained much outside help because existing political cultures of opposition facilitated the formation of broad-based coalitions.

The case in Plachimada reveals how struggles can utilize political cultures to gain popular support. Similarly, in Delhi the resistance drew on the organizing skills of trade unionists and university students who already have a strong culture of political participation. To attract more members of the general public the campaigns also capitalized on people's negative experiences with electricity privatization. Residents' experiences with the electricity privatization in the early 1990s motivated many people to join the movement. On a hot summer day, a Delhi resident passionately told me:

> [U]nfortunately the experience in Delhi has been that electricity privatization has been a failure. The service has greatly deteriorated particularly in those areas served by Reliance. I am in one of those areas. Service is atrocious! That's one more reason why people didn't want water privatized.

Many of the city's residents maintain that the privatization of the public electricity utility resulted in higher tariffs, more power cuts, and a general decrease in the quality of service. Members of the anti-privatization movement utilized people's bad experiences with electricity services to stress the problems with privatization. The populace feared that the water utility would be privatized in the same manner as Delhi's electricity utility (Karat et al, 2005). The campaigns persuasively challenged the Delhi *Jal* Board's claims that water provision would improve if a private company managed the utility. Drawing on people's experiences and existing cultures of opposition helped the movements form broad-based networks that spawned tremendous public attention and opposition to Coke's presence in Plachimada and Delhi's Water Supply and Sewage Project. All of the features of the two movements discussed thus far – cultural understandings, popular idioms, horizontal networks, and negative experiences with privatization – are elements of PCOs which Foran states explain the emergence of new movements. Foran's model closely matches the anti-privatization struggles and is useful because it elucidates their key components.

Connections: the local-global nexus

In Plachimada and Delhi global connections also propelled the movements into the spotlight and enabled them to pressure government officials to ensure people's right to water. International activists drew extensive attention to the anti-Coke movement, and campaigners in Delhi used international experiences with water privatization to emphasize the problems with such schemes. As a result, the campaigns educated the public about the local-global nexus – how the local movements related to global struggles against neoliberal globalization, US imperialism, and water privatization. Campaigns introduced global discourses against privatization and globalization and combined them with local understandings to develop a language of resistance that fueled the movements.

Medha Patkar's involvement in Plachimada exemplifies how global connections expanded the local struggle. As a leader of the National Alliance of People's Movements (NAPM), she used the Adivasi anti-Coke movement to demonstrate the problems of globalization. Patkar is a famous activist in India and within the global anti-globalization movement. In February, 2003, the activist began an anti-globalization *yatra* or pilgrimage throughout India. She started the journey in Plachimada to draw attention to the anti-Coke movement. Patkar and thousands of supporters marched from Plachimada to Trivandrum, Kerala's capital, to demand that the government take action and assist Adivasis. In her anti-globalization campaign, Patkar commonly referred to globalization as 'economic terrorism' and cited Plachimada as an example of how companies destroy environmental resources and livelihoods in the name of profit. Due to Patkar's involvement, the movement received a tremendous amount of media coverage and became known as an anti-globalization movement.

Vandana Shiva is another well-known activist, who participated in the anti-Coke struggle, and emphasized yet another dimension of the movement – its opposition to water privatization. On January 21 and 22, 2004, Shiva along with members of Plachimada's Solidarity Committee invited activists attending the World Social Forum in Mumbai to visit nearby Plachimada where they arranged the World Water Conference to highlight the problems with Coca-Cola's operations. Global water activists including Maude Barlow and Jose Bove attended the meeting and rallies against Coke. Adivasis had always complained that Coke was stealing the community's water, and outside activists used the term water privatization to describe what was happening in Plachimada. The groups released the *Plachimada Declaration* in which they asserted, 'Water is not a private property. It is a common resource for the sustenance of all' (Shiva, 2004, p.15). The protesters advocated that the government should recognize water as part of the commons, promote local control of resources, and facilitate the democratic management of water (ibid). Shiva and the international actors at the World Water Conference characterized the conflict in Plachimada as part of a global struggle against neoliberal globalization and water privatization carried out by transnational corporations.

The campaigns linked the local conflicts to struggles against the private control of water resources around the globe. Following Patkar and Shiva's involvement, the political cultures of opposition espoused by the global anti-privatization and anti-globalization movements brought waves of supporters. Kerala's left political parties including the Left Democratic Front (LDF) supported Adivasis in the wake of the growing popular support. Once in political office the LDF even banned the production and sale of Coca-Cola in Kerala as a symbolic gesture to support the anti-Coke movement. The culture of activism and left politics enabled protesters in Plachimada to gain assistance from powerful political players.

Protesters in Delhi also used anti-globalization and anti-privatization discourses to rally supporters. Vandana Shiva was one of the main organizers of the Citizen's Front for Water Democracy and organized an international water conference in Delhi to popularize the anti-privatization movement. In January, 2004, she arranged the People's World Water Forum in Delhi. The goal of the water meeting was to educate the public about the problems with water privatization, a common feature of loans and structural adjustment programs endorsed by organizations like the World Bank. The People's World Water Forum met for three days in Delhi and was supported by sixty Indian and international NGOs.

Beyond this international water conference, the campaigns showed water privatization as an outcome of neoliberal globalization and as an unsuccessful development strategy. The Right to Water Campaign and Citizen's Front for Water Democracy used a variety of methods to inform the public about anti-privatization water movements around the world. In its pamphlet and during public presentations, the Right to Water Campaign outlined fifteen failed water privatization schemes in the Global South. Protesters publicized how governments in Manila, Philippines; Cochabamba and El Paza, Bolivia; San Juan, Puerto Rico; and Johannesburg, South Africa, to name a few, implemented water reforms that gave private companies control of the public water utilities through concessions, ownership, or management contracts.

Ranjay, an activist with the Right to Water Campaign explained how organizers utilized the international experiences:

> The facts of the [Delhi] case are that this project is not going to work simply because it's not saying anything that can actually address problems of water in Delhi. It's just telling you of a preconceived, a pre-formulated solution, which has failed. And that's when the international sections, the international stories that we had sort of profiled and cases that we had looked at came in.

The cases, documented by transnational activist groups,[6] and used by the campaigns to educate the public about water privatization typically resulted in water conflicts between grassroots groups and the government, extreme increases in water bills, deterioration of water services, large numbers of disconnections, the spread of disease, and the premature termination of contracts and concessions between the government and private operators. Campaigners maintained that privatization was a 'preconceived' and 'preformulated solution' to water scarcity that ignored Delhi's unique environmental, political, social, and cultural conditions and was destined to fail just as it had in other countries. As noted in the theoretical section, Foran observes that common experiences of globalization are a component of PCOs. The global connections highlighted by activists helped them form PCOs which convinced people to participate in the protests.

The opposition also engaged in risk politics (Beck, 1992) as the campaigners highlighted the problems with similar World Bank-sponsored water projects. 'Through risk politics, environmental movements question the trustworthiness of agencies and institutions that handle uncertainties, attach probabilities and calculate risks and liabilities' (Dwivedi, 2001, p.26). The opposition exposed the problems associated with water privatization to challenge the authority of the World Bank and Delhi *Jal* Board. Activists successfully incorporated reports of failed water schemes into their campaigns and convinced the public that Delhi's water privatization project would likely fail to solve the city's water problems.

Conclusion

The formation of strong political cultures of opposition propelled the movements into the spotlight and helped protesters pressure Coca-Cola to close its factory and the DJB to abandon the privatization of the city's water utility. The campaigns advanced: shared meanings of water as a cultural and common resource, water as a human right, negative experiences of privatization, failed international water projects, and cultural histories of resistance that appealed to the public and enticed a variety of social actors to join the struggles. Scholars such as Bakker (2007 and in this book) caution anti-privatization campaigns against advocating for water as a human right because the rights framework does not require that water is designated as a non-commodity nor does it prevent private companies from taking over water provision. This chapter suggests that the human rights framework still has a place in the movement to establish the right to water. It provides a moral imperative and cultural understanding of water as a life sustaining resource that cannot be denied to people based on their inability to pay. Human rights along with notions of the commons, multiple meanings of water, and global connections were useful rallying points for the water struggles in Plachimada and Delhi. Both water movements tapped into and engendered political cultures, which helped the opposition gain vibrant and broad-based backing.

The aim of this chapter is not to champion state-led water provision as the best alternative to the privatization of resources. Rather the focus on PCOs illustrates that cultural opposition can foster forceful movements that generate democratic participation and cultivate the political will to recognize the right to water. PCOs offer opportunities to build context-specific discourses and strategies that include global repertoires such as anti-privatization and human rights. This enables movements to create languages of resistance that address the unique cultures, politics, geographies, histories, and economic realities of communities while acknowledging the connections between local struggles and global economic and political structures. If activists can tap into public sentiment, relate struggles to global movements, and define clear goals, powerful broad-based alliances can emerge. With strong local support and international attention, campaigns can pressure government officials to meet people's demands. When people harness PCOs to facilitate popular water movements, they can create the political space that allows for alternative forms of water governance and provide a path for a new global movement for the right to water to prevail.

Notes

1 The quote is cited in an anti-Coke campaign pamphlet entitled 'Coca-Cola Quit Plachimada, Quit India'. Mylamma, who was a powerful activist within the struggle, passed away in 2007.
2 Coca-Cola's wholly-owned Indian subsidiary, the Hindustan Coca-Cola Beverages Private Limited (HCCBL), retains control of the Coca-Cola bottling plant in Plachimada, Kerala. Throughout this work I refer to HCCBL as Coke or Coca-Cola.
3 Adivasis are one of the most socially and economically disadvantaged groups in Kerala and India. Adivasis are also known as indigenous people, tribals, and members of scheduled tribes.
4 This chapter draws on my dissertation research and eight months of ethnographic fieldwork on water struggles in India. To represent the variety of people involved in the conflicts, I utilize data from forty-five semi-structured interviews. Interviewees included local residents and farmers, anti-privatization activists and their pro-

privatization opponents, high ranking government officials and employees of transnational corporations, community organizers, scholars, and former Coca-Cola workers. I collected all data during two research trips to India – the first from July to September in 2006 and the second from April to September in 2007. I use pseudonyms to protect the anonymity of interviewees.

5 Water privatization is a widely contested term, but the opposition in Plachimada and Delhi maintained that the private control of water supplies constituted privatization. For a discussion of different definitions of water privatization and categorizations of water schemes, see Gleick et al (2002), Bakker (2003, 2007), and Conca (2006).

6 The campaign gathered this information from a variety of online reports by organizations including Public Citizen, the Polaris Institute, Public Services International Research Unit, and Act Against War.

References

Agrawal, A. (2005) *Environmentality: Technologies of Government and the Making of Subject*, Duke University Press, Durham, North Carolina

Alvarez, S. E., Dagnino, E. and Escobar, A. (eds) (1998) *Cultures of Politics, Politics of Cultures: Re-Visioning Latin American Social Movements*, Westview Press, Boulder, Colorado

Bakker, K. (2003) 'Archipelagos and networks: urbanization and water privatization in the South', *The Geographical Journal*, vol 169, no 4, pp328–341

Bakker, K. (2007) 'The "commons" versus the "commodity": alter-globalization, anti-privatization and the human right to water in the global South', *Antipode*, vol 39, no 3, pp430–455

Baviskar, A. (2007) *Waterscapes: the Cultural Politics of a Natural Resource*, Permanent Black, Delhi

Beck, U. (1992) *Risk Society: Towards a Modernity*, Sage Publications Inc., London

Coles, A. and Wallace, T. (eds) (2005) *Gender, Water, and Development: Cross-Cultural Perspectives on Women*, Berg, New York

Conca, K. (2006) *Governing Water: Contentious Transnational Politics and Global Institution Building*, MIT Press, Cambridge

Crow, B., Lindquist, A. and Wilson, D. (1995) *Sharing the Ganges: the Politics and Technology of River Development in South Asia*, Sage Publications Inc., California

Crow, B. and Sultana, F. (2002) 'Gender, class and access to water: three cases in a poor and crowded delta', *Society and Natural Resources*, vol 15, pp709–724

Dwivedi, R. (2001) 'Environmental movements in the global south: issues of livelihood and beyond', *International Sociology*, vol 16, no 11, pp11–31

Foran, J. and Reed, J. (2002) 'Political cultures of opposition exploring idioms, ideologies, and revolutionary agency in the case of Nicaragua', *Critical Sociology*, vol 28, no 3, pp335–370

Foran, J. (2009) 'From old to new political cultures of opposition: radical social change in an era of globalization', in K-K. Bhavnani, J. Foran, P. A. Kurien and D. Munshi (eds) *On the Edges of Development: Cultural Interventions*, Routledge, New York

Franke, R. W. and Chasin, B. H. (1994) *Kerala: Development through Radical Reform*, Promilla, Delhi

Franke, R. W. and Chasin, B. H. (2000) 'Is the Kerala model sustainable? Lessons from the past, prospects for the future', in G. Parayil (ed) *Kerala: The Development Experience: Reflections on Sustainability and Respectability*, Zed Books, London

Gandy, M. (2008) 'Landscapes of disaster: water modernity and urban fragmentation in Mumbai', *Environment and Planning*, vol A, no 40, pp108–130

Gleick, P. H., Wolff, G., Chalecki, E. L. and Reyes, R. (2002) 'The new economy of water: the risks and benefits of globalization and privatization of fresh water', www.pacinst.org/reports/_economy_of_water/_economy_of_water.pdf, accessed 1 April 2007

Iyer, R. R. (2007) *Towards Water Wisdom: Limits, Justice, Harmony*, Sage Publications Inc., Delhi

Karat, P., Bardhan, A. B., Biswas, D. and Roy, A. (2005) 'Privatization of water in Delhi', http://cpim.org/content/privatisation-water-delhi, accessed 29 March 2011

Ong, A. (1997) 'The gender and labor politics of postmodernity', in L. Lowe and D. Lloyd (eds) *The Politics of Culture in the Shadow of Capital*, Duke University Press, Durham, North Carolina

Parayil, G. (2000) *Kerala: The Development Experience: Reflections on Sustainability and Respectability*, Zed Books, London

Prüss-Üstün, A., Bos, R., Gore, F. and Bartram, J. (2008) 'Safer water, better health: costs, benefits, and sustainability of interventions to protect and promote health', World Health Organization, Geneva

Shiva, V. (2002) *Water Wars: Privatization, Pollution, and Profit*, South End Press, Cambridge

Shiva, V. (2004) *Building Water Democracy: People's Victory against Coca-Cola in Plachimada*, Research Foundation for Science, Technology, and Ecology, New Delhi, India

Strang, V. (2006) *The Meaning of Water*, Oxford International Publishers Ltd., Oxford

Sultana, F. (2011) 'Suffering *for* water, suffering *from* water: emotional geographies of resource access, control and conflict', *Geoforum*, vol 42, pp163–172

Swyngedouw, E. (1995) 'The contradictions of urban water provision: a study of Guyaquil, Ecuador', *Third World Planning Review*, vol 17, no 4, pp387–405

Swyngedouw, E. (1999) 'Modernity and hybridity', *Annals of the Association of American Geographers*, vol 80, pp443–465

Swyngedouw, E. (2004) *Social Power and the Urbanization of Water: Flows of Power*, Oxford University Press, Oxford

Truelove, Y. (2011) '(Re-)Conceptualizing water inequality in Delhi, India through a feminist political ecology framework', *Geoforum*, vol 42, pp143–152

Williams, R. (1960) *Culture and Society, 1780–1950*, Columbia University Press, New York

14

SEEING THROUGH THE CONCEPT OF WATER AS A HUMAN RIGHT IN BOLIVIA

Rocio Bustamante, Carlos Crespo and
Anna Maria Walnycki

The concept of the Human Right to Water has been a significant discourse in the water sector for some time, but on July 28, 2010 it achieved a higher status as the international water justice movement and the Bolivian government presented a resolution to the United Nations General Assembly, which lead to the declaration of the Human Right to Safe and Clean Drinking Water and Sanitation. This was an important victory for an international campaign that strives for more equitable water provision. Many of the ensuing debates have focused on the challenges in recognizing and implementing the principle, but there has been less discussion about what the right to water means conceptually, and how it forms the basis for policy within the water sector.

The principle of the Right to Water is the focal point of a discourse that has been used to promote certain political agendas and visions around water management and basic service provision. In order to explore the development of the discourse, we think it is necessary to look at some of the alliances that have formed between water activists and certain governments that may be considered 'progressive'. In this chapter, we consider how the theory and discourse of the right to water is translated into policies for water management and basic service provision in Bolivia. We explore some of the contradictions between the seemingly radical discourse of the Right to Water, and the reality of basic service provision in Bolivia, which appears to be much more conservative in reality.

Revising the concept of water and sanitation as a human right

We begin with a reflection on the so-called 'rights-based approach' that serves as the theoretical underpinning to the principle that water and sanitation is a human right. We start by reviewing two conceptual notions that relate to this approach, first the issue of appropriation of nature and the creation of rights, and secondly the issue of human rights, specifically when they involve access to and use of an element fundamental to life such as water. We believe that this reflection is necessary because while much has been written to date about how the principle of the human right to water (and sanitation) can be implemented, there is still some debate about the principle itself and the ideas that form the basis of the principle.

The creation of rights and appropriation of nature

There are still many societies and cultures in the world that manage and relate to water without a legal framework of rights. However, increasingly, it has become common, and in some cases desirable, to appropriate elements of nature so as to manage or exploit it more efficiently. Using the concept of rights (systems of control, property or possession) to frame and understand how cultures relate to and interact with their environment and its resources, runs the risk of overlooking those cultures and societies that are not based on the existence of any kind of right over nature, their environment or its resources. By insisting on using the concept of rights to frame, understand and label the diverse relationships that different societies have with water, little by little the notion of rights as a framework to understand how societies relate to water has been substantiated and validated on an international level (see also chapter by Linton in this book). At the same time, many social movements and organizations concerned with water justice have made the right to water the focus and ultimate aim of their campaigns. While the promotion of a rights-based approach to water provision has taken various ideological and discursive forms, it does, in essence promote the exclusive ownership of a resource, that was, until recently, considered a common resource, not specifically defined by limited to use for humans.

The paradox is that the concept of the right to water has come to shape the work of many social movements and NGOs concerned with water justice, who promote a rights-based approach and the appropriation of the commons, not only as a method of managing water scarcity but as a way of defending water resources against other water users such as industry, mining or neighboring communities. Even some of the most radical water justice movements, who may have an anti-privatization and anti-neoliberal focus, still maintain the logic that it is possible to appropriate nature through the creation of individual or collective rights over natural resources.

In essence the concept of the right to water, while often part of an anti-privatization discourse, is not incompatible with the privatization of water (Bakker, 2007 and chapter in this book), either through the institutionalization of individual private rights and also through collective private rights. It is clear that the human right to water and water rights are not necessarily incompatible. This can be seen in the case of Bolivia where the state has started to incorporate water rights as part of the process of institutionalizing the human right to water. The problem is that this can lead to sectionalism, in which each group seeks to protect 'their' right and to some extent, their ownership of water. This can affect the complex systems of water management based on practices of mutual aid developed by communities, which are not necessarily focused around rights. The clearest example of this situation can be seen now in the irrigation sector in Bolivia, which is where we are formalizing the rights of those who already have water without question of equity and sustainability, creating conflicts with other sectors (mainly domestic consumers) (Bustamante, 2007; Crespo, 2006).

One of the most frequent criticisms made concerning a human rights-based approach to water is that it is based on the abstract and accepted idea of what it is to be human: 'human rights cannot refer to a universal subject' (Castilla Massó, undated, p6). Such an abstract and universal understanding of what it means to be human means that the notion of a 'human right' can be somewhat vacuous. Human rights begin to take shape and become useful through the participation of politically marginalized groups. Demanding political inclusion is necessary for one to be recognized as a human capable of accessing human rights, and in turn giving substance to the notion of human rights, making them more than 'a useless formal guarantee' (Castilla Massó, undated, p6). By reframing the political landscape in which rights are formed, and by giving marginalized groups a voice in the process, Rancière suggests that there is more of chance for human rights to work to redress inequalities, in this case, for improved provision of water and sanitation (Rancière, 2006). In reality there are specific conditions required so that this can happen, and more often than not the political space for such participation is limited. In this context human rights are susceptible to becoming more akin to humanitarian rights, as states set out to establish and guarantee 'the rights of those who cannot enact them, the victims of the absolute denial of the right' (Rancière, 2006, p8). The creation of an 'other', framed as 'victims', combined with continued political exclusion justifies the ongoing occupation of political space and for the structures of inequality to continue, so that 'somebody else has to inherit their rights' and marginalized status. This is what is called 'the right to humanitarian interference' (Rancière, 2006, p8). Rancière puts this process forward through his analysis of the international interventions of the United States; however, the following paragraphs will serve to show how such a critique is useful in exploring how different state and non-state actors employ the right to water.

Some nations and organizations have taken up the cause of all those who have no access to water and sanitation as part of a fight for 'infinite justice', in which all 'distinctions are boiled down to sheer ethical conflict between good and evil' (Rancière, 2006). In the case of water and sanitation, this is expressed in the call to save the world from diseases caused by unsafe water and poor hygiene conditions, as outlined by the Bolivian Ambassador to the UN Pablo Solón:

> Diseases caused by unsafe drinking water and sanitation cause more deaths than any war (. . .) Every year more than 3½ million people die from diseases spread by contaminated water (. . .) according to the 2009 report of the World Health Organization and UNICEF entitled 'Diarrhea: Why children continue to die and what you can do': every day 24,000 children die in developing countries by preventable causes such as diarrhea water intake contaminated. This means that a child dies every three seconds . . .
>
> (Speech delivered by Ambassador Pablo Solón of Bolivia
> to the General Assembly of the United Nations in
> New York on July 28, 2010)

From this perspective the actors defending and avenging victims, legitimize interventions in order to defend the rights of 'others'. Castilla Massó suggests that this is a method employed to reinforce political order. By reducing those groups who are looking for equality to 'victims', the inevitable consequence is that 'their rights become the rights of the "other", which have to be defended in their name, so the excluded are locked into the third person, making it impossible to access the public (political) sphere' (Castilla Massó, undated) and to claim rights with their own voice.

Locking marginalized groups into the third person serves to appropriate representation in a non-democratic manner. A similar logic can be applied to the environment, which has increasingly become objectified so that it can be defended since it is not represented nor does it have any rights (except for the constitution of Ecuador, adopted in 2008). In this way the anthropocentric nature of human rights is revealed because it 'only confers rights to those who demand obligations, as a result we will not be able to grant rights to nature/the environment nor to future generations' (Boaventura de Souza Santos, 2004, p6).

The rights-based approach

Based on the reflections in the preceding sections a 'rights-based approach' has been developed with the following central tenets:

- The recognition that every person is entitled to certain rights, which are inherent to being human; this presupposes principles of equality and non-discrimination.
- The challenge is not solely to meet people's basic needs but to build and employ effective rights.
- The existence of a right also means obligations.
- The enforceability of the right by the authority in the State.

This approach can be subjected to many of the criticisms we have mentioned above, but also has some specific issues that we reflect on. Firstly, a rights-based approach has an emancipatory potential, which requires a range of conditions for this to be realized, in practice this usually involves the 'depoliticization' (Rancière, 2006) of social struggles striving for equality, as campaigns of resistance to domination make way for claims of rights. In this way, the 'political' makes way for policy and public politics, the management and administration of the social. The famous water movement in Bolivia that emerged during the Water Wars of Cochabamba provides us with an example of this. The movement formed in resistance to the domination of Cochabamba's water supply by the private sector, namely Bechtel, while its current focus is on public policy and how to improve water and sanitation coverage in the region. As a result, the technical discourse of the programs, projects, indicators, has replaced the political discourse of the Water Wars.

If we are to consider how rights can be recognized and employed by the state, we are recognizing and justifying the state as responsible for ensuring compliance. A rights-based approach means that other institutional and organizational forms are not recognized, even though they may occupy spaces for interaction and rights that don't necessarily originate from the state. As Rancière outlines in his criticism of Arendt 'it is only if you presuppose that the rights belong to definite or permanent subjects that you must state, as Arendt did, that the only real right are the rights given to the citizens of a nation, by their belonging to that nation, and guaranteed by the protection of their state' (Rancière, 2006, p7). This is implied in the case of a human rights-based approach to water: 'A human rights approach to water and sanitation provides the legal framework and ethical and moral imperative of ensuring universal access and equity. Ensuring enjoyment of human rights is not optional; governments are under a legal obligation to take action to ensure that every man, woman and child has access to the requirements of life in accordance with their human rights and dignity. This obligation can be used in advocacy to strengthen the political will and resource allocation necessary' (WaterAid, 2010).

In Bolivia, the construction of public institutions to manage water resources is expanding the role of the state in relation to water management, water services and sanitation. In this way Crespo outlines how 'we are witnessing a new state leadership, seeking to regulate, preserve, protect, manage, plan, but also drawing directly on natural heritage' (Crespo, 2010). Thus, the principle of human right to water serves to legitimize a new phase in the long history of the destruction of water commons by the Bolivian state (Crespo, 2010).

The discourses of water as a right

International water activism, particularly that related to NGOs, can be seen to be reproducing 'the politics of demand' (Day, 2004, 2005)[1]; the human right to water has become the focal point for collective action to lobby international forums in the sector. Since 2006 the international water activism has sought to structure partnerships with governments that support the principle of the human right to water. The following section will explore the alliances between activists, water-focused NGOs and governments, which were forged at the water summit in Istanbul in 2009, and who have lobbied for the passing of the UN resolution on the Human Right to Water and Sanitation.

In 2003 at the World's Water and Environment Ministers Meeting in Kyoto, which was held to discuss the global water crisis, activists of the alternative water movement protested inside and outside of the official forum. The meeting was described as being 'unrepresentative, illegitimate and opaque' (Bakker, 2007, p431). The ties of co-organizers (The Global Water Partnership and World Water Council) to private water companies and international financial institutions were also subject to criticism. The movement's demands included the Human Right to Water, along with the removal of the private sector, the return to the local water democracy, and the rejection of large dams (Bakker, 2007).

During the World Water Forum in Mexico 2006, the Joint Declaration of Movements in Defense of Water expressed firm opposition to all World Water Forums, not only for being 'areas where large transnational corporations, international financial institutions . . . and governmental powers of the world meet' (VVAA, 2006), but also for being 'exclusive and undemocratic'. It also claimed water to be a common good, access to water to be a basic human right, and that water management should be public, social, communitarian and participatory. It demanded the exclusion of water in the WTO agreements and other bilateral and multilateral treaties (VVAA, 2006).

In Istanbul in 2009, international water activists and NGOs continued discussions around similar issues and strategy development. The final declaration of the People's Water Forum outlines an intent to 'discredit the false and business-related World Water Forum' (VVAA, 2009).[2] Central themes and demands remained the same: the human right to water, the end of the privatization and commodification of water, and the condemnation of the environmental impacts of large-scale water infrastructure projects, particularly dams on marginalized and vulnerable communities (Water Alternative Forum Declaration, Istanbul, 2009).

In Istanbul in 2009, the alternative water movement, alongside allied governments, worked hard to lobby the official forum for the recognition of water as a human right. One indicator of the success of this partnership was the increase in the number of countries who signed the document, increasing from 4 in Mexico in 2006 (Bolivia, Ecuador, Venezuela and Uruguay), to at least 25 in Istanbul. The meeting resulted in an alliance being formed with several Latin American governments, which are considered to be progressive. The ministerial authorities of Bolivia, Venezuela and Ecuador participated in the panels at the alternative water forum and were greeted with applause and speeches of support. However, at the same time, it is apparent that these governments pursue domestic water policies that contradict the statements signed.[3]

The alternative forum in Istanbul brought with it a clearer trajectory than the one envisioned in Mexico 2006. According to Davidson-Harden et al (2007), sympathies and collaboration with the governments considered 'progressive' could be considered as being part of a battle of anti-capitalist positions (Davidson-Harden et al, 2007, p5). This assumes that the central conflict in this period was between the agenda of the transnational corporations and the developing opposition embodied in the human rights-based agenda, lobbying for the human right to water as part of the global commons (Davidson-Harden et al, 2007, p5). From the perspective of citizenship, the objective of the alternative water movement is 'the democratization of the state', as it 'seeks to subordinate the public institutions to social control exercised by citizens in different ways, but mainly in the search of what we call substantive exercise of their citizenship' (Castro and Lacabana, 2005).

This type of analysis, where the principal concern is the state (beyond its political or contextual relevance), minimizes the importance of the movements and collective actions that aim to reduce or eliminate relations of domination that are often racialized. Movements based around the common good and mutual aid, build and experience alternative forms of water management that are not mediated by the state or 'use' the state for their own purposes. This proposes that water is part of the commons. Such practices can be seen in the cooperatives and community-managed systems that operate in developing countries with a strong indigenous presence, where the capacity of state to intervene and regulate is weak as is the case in Bolivia.

'We all want rights'

Intellectual critics of neoliberalism and the privatization of water have proclaimed the emancipatory and democratic potential of the discourse of water as a human right. To José Esteban Castro, the human right to water constitutes part of the demands of citizenship. Citizenship rights, including the right to water would constitute an 'emancipatory vehicle under the conditions that characterize the capitalist system' (Castro, 2006, p269). They are part of the contemporary social struggles geared towards the opening, expansion and conquest (or re-conquest) of social territories delineated by existing citizenship systems (Castro, 2006, p269). Davidson-Harden et al (2007) defines the water justice movement as an anti-establishment strategy in the critique of capitalism. The movement promotes the concept of water as a human right and the global commons as opposed to the concept that water is a commodity (Davidson-Harden et al, 2007, p2). Citizenship and water hegemony are consistently constructed in relation to the state, whether that is as part of the demand for the fulfillment of its role, to criticize its decisions, or to legitimize them.

The incorporation of the human right to water as part of the international and Bolivian water activist agenda is directed towards seeking to improve the practices and functions of the state, the corporations, and everyday life with respect to the access, use and availability of water and water services through influencing the power of the state to achieve its full potential. This is in line with the politics of demand (Day, 2004, 2005). On the other hand, the rights relate to the state, in fact, the focus on economic and social rights seeks to protect and promote these rights before the state, as suggested in the term 'demand for rights'. Thus, a question emerges about how social movements can incorporate a state-lead claim such as the human right to water and sanitation into their agenda.

A characteristic of social movements and of new forms of anti-globalization activism is their autonomy from the state (Hardt and Negri, 2000). They do not want to take power or strengthen the state but reduce all forms and relationships of domination, specifically the state-related ones. In Bolivia, the long history of indigenous struggle has been to defend their autonomy and to fight against the various forms of domination that have been established, rather than to demand rights. In some cases, during the negotiations with the state they have sought 'reciprocity agreements'. Historically local communities have organized and become responsible for water management in Bolivia, particularly in rural and peri-urban areas; the state has not played a notable role in the management of water in these areas. Community management was affected as the private sector was introduced into the provision of potable water services and sanitation. This was to become the main impetus for the mobilization of irrigators, indigenous and rural groups across the country, who stood up to defend their rights to water systems and their systems of self-management.

In Bolivia in 2000, the human right to water was not part of the original agenda of Cochabamba Water Movement that emerged during the Water Wars. As explained by a key leader of the movement, Oscar Olivera, '. . . the human right to water is an issue that had not been considered here in Bolivia, even during the Water War, because . . . the movement understood water as common good, and was tied by (opposition to) to super high service fees and the expropriation of the collective effort of committees, etc., and obviously opposed to a privatization process that was cruel, corrupt and blind, where people were ignored, as if we did not exist' (Oscar Olivera, Coordinadora del Agua, personal communication, June 2010).

The fight for water in Bolivia was to halt the domination of community managed water sources that were being legitimized by privatization and commoditization of water; it was not a fight to demand for the right to water. This campaign emerged during the Kyoto Summit in 2003, which was attended by prominent leaders of the Cochabamba Water Coordinadora, irrigators, and water cooperatives from the city of Santa Cruz, FRUTCAS (Federation of Peasant Workers of the Southern Altiplano) and Latin American water activists (specifically Uruguayan activists who months earlier approved the human right to water via a referendum).

Olivera recalls the proceedings: '(In Kyoto) the discussions started around this issue (the right to water) but from a more urban perspective. They were talking about access in impoverished areas, and about how the management of water should be strictly public to ensure the "right" . . . well, I did not give much emphasis on the matter. Activists from the North, where there is a much stronger relationship between the state and civil society, drove the idea of the water as a right and the idea to lobby with the politicians, actions, that I always refused to participate in' (Oscar Olivera, Coordinadora del Agua, personal communication, June 2010).

From Olivera's criticism we note an urban vision that the human right to water is state-centered, pushed by the international activism. Apart from what has already been said around the matter of the state, it is noteworthy that an urban view of water as a right obstructs one seeing the connection with the issue of water rights. In Bolivia, at least, there is a close relationship between the urban view of water and water rights, especially in rural and peri-urban areas where most of the systems have been financed and built by the users themselves, making them 'rights holders' and not mere consumers of a service.

This brings one more element into the discussion about the relationship between water and sanitation as a human right. There is a need to distinguish between two dimensions of the right: one which is directly related to access to water of sufficient quality for human consumption; and, secondly, a more complex and wider dimension that not only recognizes the right of access to water and sanitation but also the right to participate in the decision making processes around the resource itself (water sources) and the provision of services.

In the case of Bolivia it is important to consider both dimensions, bearing in mind that as stated by Humberto Gandarillas (German Cooperation Agency Program for the Development of Sustainable Agriculture – PROAGRO, personal communication, April 2010) that over 90% of the sources of water (at least in the west of the country) are being used or are considered as part of a communitarian territory. It should be noted that the idea of a human right to water that has been individualized stands in contrast to the practices of collective rights and management that are predominant in Bolivia. The coherence of the ways and means to achieve the objectives is a topic discussed below.

The Right to Water in Bolivia

While the well-documented Water Wars of Cochabamba became the poster child and impetus for the international Anti-Privatization and Right to

231

Water Movement throughout the 2000s, in Bolivia social organizations have continued to fight various campaigns to defend water resources. FRUTCAS led a Defense of Water campaign to protect the water of local communities from export to Chile soon after the Water Wars (Quisbert, 2007), and more recently, a campaign to protect subterranean water supplies against exploitation and contamination by Japanese-owned San Cristobal silver mine (Morán, 2009). Of late, various communities and social organizations have been engaged in campaigns to defend their supplies against state-led extractive industries and hydroelectric projects, which are depleting and contaminating water supplies (Bolpress, 21 April 2010).

The human right to water has been the framing discourse for water politics in Bolivia since the election of the President Morales and the Movement Towards Socialism (MAS) government in 2006. The impetus for this came from social organizations and movements, who lobbied for the creation of the Ministry for the Environment and Water, and continued during the Constituent Assembly meetings of 2006–2007. Social organizations in Bolivia then presented proposals and discussed what responsibilities the state should have around water and its management. In these meetings water was prioritized for consumption and production, and that water sources and supplies should also be protected. It is for these reasons that the right to water in the PCS (Plurinational Constitution of the State) is outlined as follows: 'Water is a fundamental right for life.' The approved PCS of 2008 outlined the right to water for life amongst the following declarations in relation to water:

- Water is a fundamental right for life (Articles 20, 373, 374)
- The protection of water sources against contamination (Articles 374, 376)
- Respect for traditional uses and customs around water (Articles 374, 375)
- The elimination of water concession and prohibition of the privatization of water services (Articles 20, 373)
- Regulation and management of water resources with social participation (Article 374)
- Protection of water sources from free trade agreements (Article 77)
                                        (Plurinational State of Bolivia, 2008)

The government's commitment to improving access to water is outlined in Bolivia's National Plan for Development 2006–2010, and more explicitly in the National Plan for Basic Sanitation 2008–2015 which outlines a target to ensure that 90% of the country has access to water by 2015 (Water Ministry, 2008). The ensuing policies have yet to explain how the quality or consistency of water supplies will be guaranteed, particularly given that many regions of the country are living with water shortages and droughts. The MMAyA (Ministry for the Environment and Water)[4] continues with the process of institutionalizing the discourse while proposals for a new water law are in the early stages of consultation amongst social organizations and water providers.

The reality of water provision in Bolivia is that small-scale community-led and private organizations continue to dominate water provision in rural and peri-urban areas. In the rapidly expanding Southern zone of Cochabamba, the public sector provider SEMAPA does not operate. Community-led associations, water committees, and private water vendors continue to serve the majority of the region with limited financial resources and supplies, meanwhile communities continue to wait for the long-promised water from the Misicuni Dam project to arrive.[5] While the PCS explicitly states 'It is the role of the state to manage, regulate, protect and plan for adequate and sustainable use of water resources' (PCS, 2008), in the south of the city, the government has yet to demonstrate sufficient capacity, nor has it invested sufficient resources to be able to engage, or regulate the informal sector that serves this region. A system of licenses has given community-managed systems the right to use specific water sources; however many of the subterranean sources that they rely on are saline or contaminated (Ghielmi, Mondaca and Lujan, 2008). Many of the committees are fiercely independent, and are not willing for the state to take control of their water sources, instead they are looking for financial, technical or organizational support to improve access to and to ensure and protect the quality of their water supplies. For this reason, many groups continue to feel failed by the state and the policies that have emerged from the right to water discourse.

The human right to water has been a useful discourse for some groups, who have used the concept of uses and customs,[6] as defined in the PCS to protect their water sources. The 'Regantes' (irrigators) of Cochabamba successfully campaigned for a legal framework in 2006 to defend and protect their individual systems, which led to the passing of the law 3351; however, it has been critiqued by some (Bolpress, 30 July 2009) as being divisive and has been seen to cause conflicts between various user groups.

The law has introduced a system of registration, which recognizes exclusive rights for irrigators that give them to access water supplies and systems. Various critiques of the law have been raised. Firstly it could be interpreted as favoring water access for the Regantes, a group of a relatively privileged socio-economic class (Perreault, 2006), and that it does little to protect the poorest. Secondly the law does not serve to protect against the monopolization of water supplies, while the state has been critiqued for not having the capacity to intervene in such matters. This was recently exemplified in 2009, when the Coordinadora del Agua and representatives of the peri-urban water association ASICASUDD EPSAS denounced the sale of water by Regantes to peri-urban groups, and highlighted the conflict that was arising between different user groups. 'In these and other conflicts, the state is clearly absent; its incapacity to manage the situation . . . we are sure is generating similar conflicts nationally' (Bolpress, 30 July 2009).

There is an assumption within the notion of 'uses and customs' that those who have a registered system are best placed to protect their water supplies, whereas often many of the Regantes pursue farming practices that are unsustainable, potentially polluting, and beyond the remit and intervention of the state (Bustamante, 2007). While the international discourse of the human right to water focuses on the state's obligation to provide water to individual citizens, in the Bolivian context that has a rich history of community-level resource management, the process of institutionalizing the right to water has attempted to engage and incorporate come of the traditional systems of water management. The challenge in this approach is to avoid sectorialism and conflict between different user groups.

The capacity of the PCS to protect water sources and the environment

The human right to water in Bolivia was established as part of a spectrum of social, political and economic rights in the PCS, which should protect and empower historically marginalized communities. Meanwhile the Bolivian government is pursuing a program of ambitious industrialization through The Great Industrial Leap,[7] therefore continuing to prioritize water for use in the extractive industries. This has lead communities to call into question the PCS, and question who really has the right to water in Bolivia, and whether or not introducing a rights-based approach to water can protect and provide for marginalized groups, or whether the same inequalities and injustices that were observed under neoliberalism continue through The Great Industrial Leap. The PCS contains a range of overlapping and supportive rights that could be employed by communities to defend their water supplies and protect against environmental injustices. While the Right to Water may be one element, the Right to Consultation over large-scale extractive projects, is another that could be employed. In practice, as we shall see in the case of the Corocoro copper mine, such rights have been sidestepped or ignored by the Bolivian government.

The Corocoro open cast mine is located in the Provincia Pacajes in the department of La Paz, some 140 kms outside of La Paz. It was Bolivia's biggest copper mine until 1985, when an international slump in the price of metal forced it to close. In August 2009, after receiving $19 million worth of investment from the Bolivian government it was inaugurated by President Morales. This was part of a joint enterprise between the Bolivian state mining enterprise COMIBOL (The National Mining Corporation of Bolivia) and Kores, the South Korean state-led mining company. Having recently completed its first year, it has produced 3,008 tons of copper, to the value of $2.6 million. While there has been support for the mine and the employment and economic growth it has brought to the region, the community of Jacha Suyu Pakajaqi as represented by the social movement CONAMAQ (Consejo

Nacional de Ayllus y Markas de Qullasuyu) have sought to use the PCB and associated laws to protect their water supplies and local environment (CONAMAQ, 2010). Their plight highlights how the rights enshrined in the PCS and associated laws have failed to protect the water sources and local environment of Jacha Suyu Pakajaqi.

Since the mine re-opened, Jacha Suyu Pakajaqi have anecdotal evidence of decreased water levels and changes in water quality, and so CONAMAQ's Commission on Extractive Industries requested the Environmental Impact Assessment for the mine, an outline of how COMIBOL would mitigate any of the mine's impacts of the local environment and water resources, and most importantly community level consultation on the mine. Article 11 of the PCS states that citizens are entitled to consultation which is: 'direct and partici-patory, through referendum, the citizens' legislative initiative, the recall, the assembly, the council and consultation'. This was then expanded on in law by the Electoral Systems Law[8] that contradicted the PCS, as was to become evident in the ensuing governmental response.

The government's response to CONAMAQ has been that they do not possess the institutional capacity to recognize their rights as set out in the PCS (Letter from the Ministry of Mining Metallurgy, October 2010), or the Electoral Systems Law.[9] Furthermore, even with consultation, the Electoral Systems Act under-mines the rights set out in the PCS, stating that while consultation should be undertaken, it does not have to influence the decisions of the government. Hav-ing been failed by the PCS and the Electoral Systems Act, CONAMAQ have recently decided to present evidence of the allegations of environmental pollu-tion from the Corocoro mine in front of the Inter-American Court of Human Rights of the Organization of American States (Los Tiempos, October 2010).

The Bolivian government's commitment to expanding the country's hydro-electric production, through the construction of several large scale hydroelec-tric plants in the Amazon, further demonstrates how large scale industry has been prioritized over the livelihoods of local communities and their environment, as part of The Great Industrial Leap. This is to serve the energy needs of local communities, but also, and more controversially to create energy for export to Brazil. As stated by the Vice President: "Evo is going to build this hydropower plant that elders did talk about, and raise the dream of our region, of our department and our country" (Vice President Álvaro García Linera, 18 October 2009).

The Basin of the Madeira River, which spans Peru, Bolivia and Brazil, has been the focus of many of Brazil's most recent hydroelectric projects. Four dams have been built in Brazil, and multiple environmental impacts have been noted from the Jirau and Santo Antonio dam, including the diminishing of fish stocks, and increased levels in malaria. In 2007, the Brazilian Institute of the Environment and Natural Resources (IBAMA) recommended that a more detailed Environmental Impact Assessment be undertaken, in order to have a more detailed understanding of the impacts, but the Brazilian government have failed to act on this. It is suspected that these hydroelectric projects are impacting on the fish stocks of downstream rivers of Bolivia, which contains 80% of Bolivia's commercial fish stocks. The Brazilian government has been advocating for the promotion of hydro-electric projects by the countries that share their river basin, with a view to energy export to Brazil (Molina, 2010).

In 2009, the Bolivian government invested $8 million for the planning and design phase of a hydroelectric project on the border with Brazil in Cachuela Esperanza. The Canadian consultancy Tecsult-Aecom are undertaking the project, along with scoping exercises to explore the viability of hydroelectric projects on the rivers Madeira, Mamore and Beni, as well as to explore the environmental impacts of Jirau and Santo Antonio on Bolivian territories. This project is part of a larger commitment by the Bolivian government to invest in hydroelectric power as one of their pillars that makes up The Great Industrial Leap; if approved, work is likely to start in 2011, while the project itself would come into operation in 2020 (Bolivian Ministry of Hydrocarbons and Energy, 2010). However, the long-term impacts of such projects are concerning, the environmental impacts of dams, including increased inci-dences of flooding and malaria, are well documented, particularly in the Amazon basin, but beyond this the Bolivian government would be wise to take note of the experiences of their Brazilian neighbors who share the basin; diminishing fish stocks and increased incidences of malaria. Furthermore there has been no information around how local communities may be af-fected, or, consider the experiences of the communities surrounding Corocoro, whether they will be consulted with.

The need for reflection

There exists a demand for new paradigms that could inform lasting solutions to water and sanitation provision, and so any initiative that is oriented in this direction brings with it a level of expectation. This has been the case with the principle that water and sanitation are a human right. The interna-tional water justice movement has been keen for the principle to be adopted on an international level, and continue to be dedicated to making sure it is implemented in the future. However, and as some have already pointed out, it is necessary to look beyond the debate, to explore how the principle can be achieved, what its theoretical foundations are, how the discourse can be constructed and advanced, and also to learn from concrete experiences of implementation. Only then will we see the full picture.

In the case of Bolivia, discussed briefly in this chapter, understandings of the human right to water are complicated by the fact that the concept is not dissimilar from pre-existing understandings of water rights, especially those recognized as systems under the notion of 'uses and customs' in rural and peri-urban communities, where the state has, for the most part, been absent. In this context, to fully develop water as a human right would also develop the state's role as a provider and guarantor of this right, that is, to legitimate it. On the one hand this gives the state the power to undertake something that historically it has been unable to do: to define, then grant or deny these rights under their own terms and laws. This may seem logical in other countries, but this is not the case in Bolivia, where the autonomous management of resources has long operated without state intervention.

Under the principle that water is a human right the state acquires the ability to intervene on behalf of those without water, the 'victims' (Rancière, 2006), who are unable to claim for themselves and need to be represented as political subjects in order to achieve or exercise their rights. Thus the cause for the right to water and sanitation becomes a struggle between 'good and evil', in which everyone must put aside their interests and differences for the success of 'collective good', so that everyone can receive water and sanitation. This involves a process of depoliticization (Rancière, 2006) of political subjects and of the cause itself, leading to a process in which political debate is diluted and becomes a public policy debate, managed by the state, in a framework defined by the state, where there is very little room for dissent or participation.

Dissent still exists and manifests itself in various forms, as we have seen in Bolivia, and it is gaining momentum through resistance to a series of conflicting public policies. This highlights the contradictory practices that exist in a country where water is a human right but which affects water rights and pre-existing systems for managing water resources in rural, peri-urban areas. The acts of resistance and the defense of local ways of living, as observed in Bolivia, exemplifies the need to reflect more deeply on the significance of the right to water as a discourse, what it means in practice, and what that means for the direction of the continued campaign for water justice and equality in water provision.

Notes

1 Richard Day (2004, 2005) and Harrington (2010) distinguished the politics of so-called new social movements as being between politics of demand and politics of the act. The latter, rejects the idea that the state will improve the living conditions of certain groups and social sectors, and is also unwilling to interact with the state, their practical politics employ experimental, creative and pre-figurative action. The politics of demand aims at improving the practices of states, corporations and everyday life. Influencing or using state power to achieve effect, these practices demand rights from the State or other agencies such as international cooperation.
2 In fact, the document reaffirms 'all the principles and commitments expressed in the Declaration of Mexico City 2006.'
3 A week before the World Water Forum in Istanbul, the government of Ecuador (a country that has included the human right to water, and the rights of nature as part of their constitution), took away the operating license of the NGO 'Ecological Action'. The NGO had criticized the government's mining and environmental

policy and mining, specifically its impact on poorer, specifically indigenous communities. This is hardly justifiable within the international water justice movement.

4 The former Water Ministry became the Ministry for Environment and Water (MMAyA) in 2009.

5 Misicuni is a three-phase project that will channel the water of up to three rivers to the city of Cochabamba and its surrounding area. It has been in development since the 1960s and it's unclear as to when the work will be completed, or whether appropriate infrastructure will be constructed to bring water to the southern zone of the city.

6 'Usos y Costumbres' meaning 'Uses and Customs' in relation to water refer to the customs and traditions by which water has traditionally been managed, predominantly in rural settings. These have been protected by the constitution in Bolivia.

7 The Great Industrial Leap or El Gran Salto Industrial in Bolivia is the term for the MAS-IPSP government's program to invest in the industrialization of its natural resources to promote economic growth in Bolivia.

8 Article 39 of the Electoral System Act (Ley de Regimen Electoral) states 'Prior consultation is a constitutional mechanism of direct and participatory democracy, convened by the Plurinational State of Bolivia. It is compulsory prior to making decisions about the projects, works or activities related to natural resource exploitation. The people involved will participate in a free, prior and informed manner.

In the case of the participation of nations and peasant indigenous peoples, consultation will take place in accordance with their rules and procedures.

The findings, agreements or decisions made in the context of prior consultation is not binding, but must be considered by the authorities and representatives in decision-making levels that apply.'

9 '(The) institution to undertake the consultation is not yet in place, so it is impossible for us to organize and initiate prior consultation at this time ... In conclusion, we are informing you that we will not be able to undertake the consultation process' (Letter from the Ministry of Mining Metallurgy, October 2010).

References

Bakker, K. (2007) 'The "commons" versus the "commodity": alter-globalization, anti-privatization and the human right to water in the global South', *Antipode*, vol 39, no 5, pp953–955.

Bolpress (21-4-2010) Declaración de la Mesa 18 de la Conferencia Mundial de los Pueblos por el Cambio Climático, www.bolpress.com/art.php?Cod=2010042105, accessed 21 April 2010.

Bolpress (30-07-2009) Denuncian que campesinos y regantes del MAS quieren cobrar por el agua en Cochabamba, www.bolpress.com/art.php?Cod=2009073003, accessed 16 September 2010.

Bustamante, R. (2007) ' "... y quien se hace dueño del agua del río ...": Re-pensando y de- construyendo el enfoque de los derechos sobre el agua', ponencia presentada al Seminario Internacional: 'Modelos de Gestión del Agua en ciudades y comunidades de los Andes'; La Paz del 5 al 8 de Noviembre del 2007.

Castilla Massó, J. (undated) 'Democracia, ciudadanía y derechos humanos en la obra de Jacques Rancière', Universidad Complutense de Madrid.

Castro, J. (2006) 'Agua, democracia, y la construcción de la ciudadanía'; en VVAA, La Gota de la vida: Hacia una gestión sustentable y democrática del agua La

Gota de la vida: Hacia una gestión sustentable y democrática del agua; México: Fundación Heinrich Boll; 400 pp. www.boell-latinoamerica.org/download_es/ Libro_La_Gota.pdf

Castro, J. and Lacabana, M. (2005) 'Presentación. Agua y Desarrollo en América Latina: por una democracia sustantiva en la gestión del agua y sus servicios', *Cuadernos del Cendes*, vol 22, no 59.

CONAMAQ (2010) 'Consejo de Gobierno Originario respalda demanda de Jach'a Suyu Pacajaqi', published in www.conamaq.org/index.php, accessed 12 November 2010.

Crespo, C. (2010) 'El derecho humano en la practica: la política del agua y los RRNN del gobierno de Evo Morales', CESU – CISO – UMSS Ponencia al Congreso de Sociología.

Crespo, C. (2006) 'Hacia una política de los bienes comunes del agua en Bolívia: los desafíos y contradicciones de la agenda post "guerra del agua"', ponencia presentada al taller de investigación en agua y gobernabilidad organizado por el ISF en Barcelona el 16 de Noviembre de 2006.

Davidson-Harden, A., Naidoo, A. and Harden, A. (2007) 'The geopolitics of the water justice movement', *Peace Conflict & Development*, issue 11, www.peacestudiesjournal.org.uk.

Day, R. (2005) *Gramsci is Dead: Anarchist Currents in the Newest Social Movements*, Pluto Press, London.

Day, R. (2004) 'From hegemony to affinity. The political logic of the newest social movements', *Cultural Studies*, vol 18, no 5 September, pp716–748.

de Souza Santos, B. (2004) '*Los derechos humanos y el Foro Social Mundial*', ponencia presentada en el XXXV Congreso de la Federación Internacional de los Derechos Humanos, FIDH, 2 al 6 de Marzo de 2004.

Ghielmi, G., Mondaca, G. and Lujan, M. (2008) 'Diagnostico sobre el nivel de contaminación de acuíferos en el Distrito 9 del Municipio del Cercado en la ciudad de Cochabamba y propuesta para su protección y control', en ACTA NOVA; vol 4, no 1, Artículos científicos pp51–86.

Gobierno de Bolivia (2004) Ley de promoción y apoyo al sector riego para la producción agropecuaria y forestal, Ley No 2878 de 8/X/2004; La Paz.

Hardt, M. and Negri, A. (2000) *Empire*, Harvard University Press, Cambridge.

Harrington, C. E. (2010) 'Rethinking the divide: beyond the politics of demand versus the politics of debate', submitted to the University of Exeter as a thesis for the degree of Doctor of Philosophy in Politics, 206pp. https://eric.exeter.ac.uk/ repository/bitstream/handle/10036/3143/HarrisonC.pdf?sequence=2, accessed 14 November 2010.

Los Tiempos (October 2010), www.lostiempos.com/diario/actualidad/nacional/20101020/ el-conamaq-presentara-pruebas-del-caso-corocoro-ante-tribunal-de-la_95171_ 183850.html, accessed 30 October 2010.

Molina Carpio, J. (2010) 'Es viable el proyecto Cachuela esperanza?', published 12-02-2010, www.fobomade.org.bo/art-720, accessed 2 December 2010.

Morán, R. (2009) *Minando el agua: la mina San Cristóbal, Bolivia*; Cochabamba: FRUTCAS/FSUMCAS/CGIAB.

Ostrom, E. (2000) 'Private and common property rights', Workshop in political theory and policy analysis, Center for the Study of Institutions, Population and Environmental Change, Indiana University.

Perreault, T. (2006) 'From the Guerra del Agua to the Guerra del Gas: resource governance, popular protest and social justice in Bolivia', *Antipode*, vol 38, no 1, pp150–172.

Plurinational State of Bolivia (2008) Plurinational Constitution of the State (PCS).

Plurinational State of Bolivia (2010) Ley de Régimen Electoral.

Quisbert, F. (2007) *Proteger y preservar las aguas subterráneas del sudoeste potosino*, Cochabamba: FRUTCAS-FSUMCAS.

Rancière, J. (2006) 'Who is the subject of the rights of men?', 16 Beaver Articles, www.16beavergroup.org/mtarchive/archives/001879.php, accessed 21 September 2010.

VVAA (2006) Declaración conjunta de los Movimientos en Defensa del Agua, 3pp. www.comda.org.mx

VVAA (2009) Declaración del Foro Alternativo del Agua – Estambul, 19 de marzo de 2009, 3pp.

WaterAid (2010) 'Human rights approach to development', Rights and Humanity www.righttowater.info/code/HR_approach.asp, accessed 23 October 2010.

Water Ministry Bolivia (2008) National Plan for Basic Sanitation 2008–2015.

Other documents

Embajada de Bolivia en Francia (2010) Comunicado de Prensa – EBFR-Nr. 39/2010, Onu Declaró Al Agua Y Al Saneamiento Como Un Derecho Humano, A Propuesta De Bolivia.

Discurso dado por el Embajador Pablo Solón del Estado Plurinacional de Bolivia ante la Asamblea General de las Naciones Unidas en Nueva York, el día 28 de julio de 2010. http://cmpcc.org/ 2010/07/28/discurso-derecho-humano-al-agua-y-saneamiento/, accessed 3 November 2010.

15

FROM COCHABAMBA TO COLOMBIA

Travelling repertoires in Latin American water struggles

Verónica Perera

In English the term 'move' has more meanings to provoke us, and these are worth exploring. To move is to travel. To be moved is to open one's heart . . . Social movements . . . grow from traveling forms of activism as well as the transformation of consciousness.

(Anna Tsing, *Friction*)

Introduction

On May 18, 2010, fourteen days before my arrival in Bogotá, the Colombian Congress rejected 'the referendum on water', as Colombians call it, after more than three years of national mobilization. Congress members refused to discuss the text of the referendum, which had been crafted by the water movement and endorsed by more than two million citizens. In addition, three days before I landed in Colombia, 'Uribe's follower' Juan Manuel Santos had won the first round of presidential elections. The right-wing coalition that, it appeared, would continue to dominate Congress for the next term was, in the words of environmentalists, 'the main enemy of the referendum on water'. In such a gloomy context, I thought, activists would be demobilized, their spirits low, and the organization weakened. I decided I would still do my field trip to research the water movement, but it would be, I anticipated, like going to a wake to offer condolences.

I was wrong. And puzzled by the high energy of the well-attended IX National Assembly of the movement for water and life at the headquarters of the water workers' union, where I heard cries like 'Ours is not a defeat' or the overenthusiastic and off-tune 'Viva el referendo por el agua!' With the 'support of international delegates' at the assembly (from Uruguay, France, Belgium, Italy and myself from Argentina included), activists reminded themselves about being a 'movement rather than a referendum' and renewed their commitment to continue the struggle. Either activists were reading the political conjuncture in a much more optimistic way than I was, or I still needed to understand what the water referendum was really about. It turned out the latter was true. Like the alter-globalization movements that Seattle inaugurated, the Colombian water movement managed to bring together 'turtles and teamsters', or environmentalists and unionists, among other actors, who coalesced around the organization of a referendum to include the human right to water in the constitution. Colombians were inspired by the iconic 2000 Cochabamba water war, when the multitude – to use Bolivian analyst García Linera's (2004) description – cancelled a privatization contract and evicted a United States-led transnational corporation. Colombians also followed the footprints of Uruguayans who, in 2004, amended their constitution to include both access to water as a fundamental human right, and mandatory supply by state-owned companies, and regulated those rights in a new water law with a high level of citizen participation in 2009. Enthused by the Bolivian and Uruguayan struggles, Colombians created the Comisión Nacional en Defensa del Agua y la Vida (CNDAV, National Commission for the Defense of Water and Life), which, besides environmentalists and unionists, also included public service organizations, Afro-Colombians, indigenous groups, women's collectives, delegates of community aqueducts, youth organizations and human rights advocates.

At that assembly in Bogotá, I began to understand a conversation I had heard a year earlier at a Red VIDA workshop at the 2009 World Social Forum (WSF) in Belem, Brazil.[1] Activists, venting frustration, were questioning the political efficacy of constitutional reforms for water justice. Diego, a Colombian environmentalist from Bogotá, trying to reframe the terms of the debate and de-emphasize the legal reform, said, 'Besides the legal change, which does not exhaust our struggle, the referendum is ultimately a pedagogic exercise, an exercise of direct democracy, and an exercise of the territories' – a statement which I elaborate upon below.[2] And Juan, a Uruguayan activist, added, 'Colombia is the most important symbolic struggle now taking place. It needs the commitment and explicit support of us all.' The fact that I had heard all this at the WSF in the Brazilian Amazon, and later found myself on a field trip to Colombia, was not a mere contingency in a multi-sited research process. Colombian activists inscribe themselves within the alter-globalization movement. 'We encountered the global movement for the human right to water at the WSF, and understood it as a good vehicle for what we were trying to do', the official spokesperson of the referendum told me on June 9, 2010.

Since the 2000 water war in Cochabamba, water struggles became prominent in Latin America, and highly visible within the WSF process. Water struggles are deeply rooted in places – or particular historical geographies, political economies and cultural contexts. Yet, since Cochabamba, there has been an *emerging activist repertoire* that travels, intertwining bodies and places, and building networks of activists and scholars, like me. In such travelling along transnational public spheres, and through 'digital networking' (Juris, 2008) that creates 'communications internationalism' (Waterman, 1998) activists circulate ideas and symbols (like the name Commission for the Defense of Water and Life); emotions, varying from rage and frustration to hope and solidarity; languages with which to frame water issues as unjust and morally wrong; information about policies and corporations; declarations; and alternatives like public-public or public-community partnerships. They build trust – in each other and in their chances of success. In doing so, they craft a glocal (Robertson, 1995) individual and collective identity around notions of water justice that becomes available for mobilization for socio-environmental agendas. These are 'activists' packages that travel', Anna Tsing says, 'that come to us in allegorical bundles' and 'are translated to become interventions in new scenes where they gather local meanings and find their place as distinctive political interventions' (Tsing, 2005, p238).

Despite the hesitation of some, and despite the fact that it was not part of the 2000 Cochabamba water war (see chapter by Bustamante et al) amending national constitutions to include the human right to water is a key piece of this travelling activist package. Bakker (2007, and in this book) cautions against framing anti-privatization campaigns in terms of 'water is a human right and not a commodity'. While 'human right' is a legal category for individuals, entitling them with rights vis-à-vis the state, 'commodity' refers to the property regime of the resource. The right to *access* water does not automatically define the character of water as a non-commodity, and thus does not foreclose the provision of water by private corporations. Water warriors should, instead, endorse the notion of the commons as the best principled and most strategic choice, Bakker contends. On the other hand, others in this book argue that 'human right' and 'commons' do not necessarily exclude each other, and are often part and parcel of the same political culture of opposition that activists draw upon (see chapter by Bywater); and if the human right to water is to be fully materialized, it needs to incorporate the logic of the commons and a focus on citizens' participation (see chapter by Clark). In addition, if we are to overcome the anthropocentric, individualistic, possibly capitalist, and state-centric nature that Bakker attributes to 'human right', Linton argues in this book, such a right needs to be revised through the lens of the collective, processual and interrelated notions of species-being and social production of wealth.

In this chapter, I join the latter voices that, even while qualifying it, still understand the human right to water as a fertile discourse and mobilizing tool. Drawing on multi-sited fieldwork, I explore the Colombian water movement and the activist repertoire that has been travelling since Cochabamba and Uruguay.[3] Beyond the diffusion of strategies (Tarrow, 2005), I examine, prioritizing activists' own reflexivity, the way in which the human rights universal engaged and became useful for local political projects from below. 'To study engagement requires turning away from formal abstractions to see how universals are used', Tsing writes (2005, p9). And she adds, 'Universals are effective within particular historical conjunctures that give them content and force' (p8). Within the Colombian referendum, the universal of the human right to water became an effective language also to defend collective life projects embedded in the idea and materiality of the territory, and to confront development based on global capital, extractive industries and the terror of the armed struggle.

Earlier stories of success

The Bolivian, Uruguayan and Colombian water struggles opened up new spaces of democratic participation, and glued together constellations of actors that had never cooperated before. Besides drawing on the symbolic capital of the name CNDAV, which by 2007 had acquired its own prestige, Colombian activists were also influenced by the flexible mobilizing structure of locally rooted actors from Cochabamba and Uruguay. Colombians paralleled the 'multitude-form' – the expression that Bolivian sociologist García Linera (2004) used to explain how preexisting territorial organizations (i.e., *regantes* or indigenous/small farmers' irrigation groups, and neighbourhood-based associations) came together as a collective for the first time in the Cochabamba revolt, after neoliberal reforms had weakened unions and the historically powerful Bolivian Workers Federation (Central Obrera Boliviana, COB). In all three experiences, place-based organizations assembled with environmentalists, contentious unionists, and leftist activists from political parties or consumer-based associations.

The 'neoliberalization of water' (Bakker, 2007 and in this book)[4] was the crucial process that prompted collective action in Colombia, although with less of a flurry than in Cochabamba, and in a political culture with fewer bottom-up opportunities than Uruguay. Let me briefly elaborate on these antecedents. As is well known, when in 1997 the World Bank included water privatization as a condition to extend Bolivia's debt relief (Schultz, 2003; Barlow and Clarke, 2004), Cochabamba's water supply management was leased in a closed-door decision to a United States-led corporation with a Spanish name, Aguas del Tunari. Because only half of Cochabamba's 500,000 inhabitants had access to public piped water and because local elites had been discussing the construction of large infrastructure to improve domestic, agricultural and industrial water provision which lacked funding, there was a somewhat favourable public opinion towards privatization. Yet the corporations' lack of investment commitments, the future rate hikes and the way in which the new legal tool would undermine *regantes* customary laws on collective water ownership and management became obvious very soon

(Tapia, 2000) and sparked organized resistance in both the city and the countryside. At the first demonstration-turned-town-hall-meeting, organized by the CNDAV in Cochabamba, people decided to demand the government repeal the privatization law, tear up the contract and reverse rate hikes (Olivera, 2004). Amidst a state of siege, the contract was indeed cancelled and the transnational corporation evicted from the country.[5] Today, Bolivia, and Cochabamba in particular, are still not free from water stress, far from universal supply and water justice. Conflicts over water sources and provision still persist among rural and city groups, and with the national state – especially within 'The Big Industrial Leap' plan sponsored by the Morales administration (see chapter by Bustamante et al). Yet, from Hollywood movies to the WSF, 'the Cochabamba water war' travelled the world over as a story of success: networked activists had been powerful enough to undo the neoliberalization of water.

In Uruguay, where 91.4 per cent of households have access to potable water, privatization began in 1993 with the partial concession of the utility company of the mostly wealthy and touristic Maldonado County. A few years later, the whole company was sold to transnational capital. Maldonado households experienced rate hikes, even if smaller than those in Cochabamba.[6] In June 2002, the government signed a letter of intent with the International Monetary Fund agreeing to change regulations and promote private investment in sanitary services (Santos, 2006). The Batlle administration aspired to extend the private sector model from the Maldonado County to the rest of the country through concession regimes, avoiding privatization laws. Because civil society organizations, over the last decades, had begun to use referendums, a legal tool of direct democracy hitherto monopolized by political parties,[7] the Uruguayan government feared that citizens might network around a referendum to revoke a possible privatization law. In addition, civil society had also set the remarkable precedent of the 1992 plebiscite that had banned the privatization of any public service.

Thus, activists understood that if the government were to privatize without passing any new law, the strategy to prevent privatization demanded a *constitutional reform*. Emboldened by Cochabamba, they formed the CNDAV, and planned a referendum to constitutionalize the access to drinking water and sanitary services as fundamental human rights to be exclusively and directly provided by state actors. On October 31, 2004, more than 64 per cent of Uruguayans voted in favour of the constitutional water amendment. Because of bilateral investment treaties protecting corporations, like in Cochabamba, de-privatization was not immediate – transnational corporations ended up leaving the country but with monetary costs for the Uruguayan state, while private national companies remained as suppliers for few more years. Yet, the Uruguayan case journeyed the world over as another story of success: networked citizens had constitutionalized the human right to water and its mandatory provision by public companies.

In Colombia: red, green, glocal

Colombian activists identify the beginning of the neoliberalization of water with Law 142 which, in 1994, re-regulated all public services – water, power, gas and telecommunications – 'with an entrepreneurial spirit', 'opening the door to privatization' (Salazar Restrepo, 2010), encouraging private investment, and introducing principles of competition and 'economic and financial efficiency mainly by taking tariffs to "real" levels' (Urrea and Camacho, 2007). The first experiment to create a market of water with the blessing of global financial institutions and transnational corporations took place in the mid-1990s in Barranquilla and Cartagena. Alerted by the Caribbean experiment, activists foresaw 'a trend that would be extended to the whole country through the Water Department Plans' unless 'networked' citizens contested it 'strategically' (Urrea and Camacho, 2007, p38). Activists and analysts claim that, en route to privatization, the first goal of such government plans since 2008 has been to create economies of scale by pressuring municipalities to aggregate water providers in the department. Municipalities thus often face disproportionately high technical requirements, and end up being decertified as legal water providers. For activists, this means a threat to the survival of community aqueducts, and the dispossession of collective wealth and public infrastructure 'built with people's monies along generations'. 'All this can be stopped with the referendum', a water worker unionist told me on June 8, 2010. And 'Colombia's hydro sovereignty' against 'chameleonic transnational corporations' can be defended with the referendum on water, activists wrote in 2007 (Urrea and Camacho, 2007, pp37–38).

The 'red green alliance' (Bakker, 2007 and in this book) thus came into being on February 24, 2007, when roughly forty organizations, already identifying themselves as a 'water movement', met at the office of the government ombudsman and decided to collect citizens' signatures for a referendum on water. At that point, environmentalists had realized that, beyond 'narrowly defined environmental projects', they needed to engage larger political transformations. They had discovered at the WSF that there was a *global* movement which framed the struggle in terms of 'the human right to water', aimed at constitutional reforms, and 'looked like a good way to go'. They were aware of their difference with Uruguay – Colombia had no history of popular referendums. This manner of participation had only been enabled by the 1991 Constitution and, according to Hugo Armando, made some on the left, and some environmentalists, suspicious. Others doubted whether the legal strategy would end up weakening the social mobilization. And for others, the rights framing had flaws: 'I know that speaking about the human right to water is anthropocentric – water is essential for *all* beings, not just humans', Diego told me on June 8, 2010.

Yet, the rights framing registered them, globally and in Colombia, within a discursive space that captured the spirit of their project. It differentiated them from opponents of global financial institutions, or, in the words of a Red VIDA member, 'the confusing if not misleading talk at the corporate-led World Water Forum', where water was defined and defended as basic need but not fundamental right. It aligned them, globally, not only with Uruguay but also with struggles that had consitutionalized such a right years later in Venezuela, Ecuador and Bolivia. Within Colombia, the rights framing partnered them with struggles seeking to defend the materiality and the idea of 'territory' in the context of displacement and terror, which I elaborate upon below, and which Samuel, a social leader from the Pacific emphasized in a conversation we had on June 18, 2010. And, as Diego put it, 'In Colombia, if you say you struggle for human rights it is clear where you stand.'

Belonging to the Latin American movement meant more than sharing the human rights frame. At critical conjunctures, world-renowned activists gave visibility and increased legitimacy to the Colombian campaign. For example, Uruguayan unionist Adriana Marquisio, Bolivian worker Oscar Olivera, Spanish hydrologist Martínez Gil and Canadian UN consultant Maude Barlow participated in innovative voyages along the rivers Atrato, Sinú, Alto Cauca, Bogotá, Meta and Magdalena. After the 1991 constitutional reform that recognized Colombia as a pluri-national and pluri-ethnic state and granted black and indigenous communities collective property in the Pacific region, activists from the Process of Black Communities (PCN in Spanish) organized voyages along rivers to recognize the territories they were begin-ning to learn as their own (Villa in Escobar, 2008). Drawing from that legacy, and with the goal of opening the conversation to include the concerns that river communities cared most about, the referendum campaign turned its 'gaze back to the river'. Freshwater became the end and the means of the campaign, and as activists journeyed, they produced networking events. Thanks to the preparatory work of the teachers' union, wherever the canoes made a stop locals welcomed the activists warmly, and they all gathered in forums (conversatorios), and at music festivals and plays dealing with water issues and everyday life by the river. Most of the time, these events left behind stronger 'territorial committees' that would later continue discussing water issues and collecting signatures for the referendum.[8]

Support from prestigious global activists was also empowering for Colombians after the first debate in Congress. Under the influence of the Uribe administration, and overriding citizen participation (i.e., more than two million signatures) and the most crucial demands of the movement (the human right to water, water as part of the commons, and the state's obligation to provide a minimum universal), Representatives proposed a very impoverished alternative text for the referendum. At that point, a trans-national coalition of environmental NGOs and unions of water workers, students, human rights and faith-based organizations from 58 countries sent a letter to the House and the Uribe administration condemning the modification of the original platform, and highlighting the significance of the Colombian referendum for the 'whole global water movement'.[9] Empowered by this letter and the local mobilization, activists appealed the decision of Congress, and prevailed.

The text

'The knowledge that makes a difference in changing the world is knowledge that travels and mobilizes, shifting and creating new forces and agents of history in its path', Tsing writes (2005, p8). Colombians inherited knowledge from their fellow water warriors in the region. But the five-points platform of their referendum text was the outcome of a national grassroots conversation that lasted more than a year. As they adapted the repertoire, and discussed the principles and rights for amending their own constitution, they created new forces, and expanded the emancipatory potential of the water struggle. At play were desires to deepen democratic participation, fuel public dialogue about 'development', redefine water as part of the commons and evoke, what scholars would call, hybrid notions of social nature.

The referendum text establishes the *access* to potable water for all as a fundamental human right. It also characterizes water, in all of its manifestations, forms, and states as a *non commodity* – 'a common and public good', 'of public use', and a 'good belonging to the nation' – specifying that waters within the collective territories of indigenous and black communities are integral to those territories, and that the state will guarantee the sacred meanings these ethnic groups give to water. The state is also obliged, according to the referendum text, to *protect all waters* 'given that [water] is essential for the life of all species and for current and future generations', and to protect 'ecosystems essential for the cycle of water', which should be prioritized to preserve this cycle.[10]

With this text, CNDAV activists, and environmentalists in particular, also aspired to bring 'development' to the fore of the discussion and challenge extractive industries, especially large-scale mining. These industries have of course a long (colonial) history in Latin America, where silver extraction from the Bolivian city of Potosí is the 'symbol of a plundering culture' that since the sixteenth century 'fed the European coffers and the early industrial revolution' (Svampa and Antonelli, 2009, p15, my translation). Since the late 1990s, however, like in other Latin American countries, the imagination of 'a mining country in a globalized world' has nourished elite discourses and governmental practices, where open, large-scale mining by transnational corporations is envisioned and sought out as a novel and desirable strategy of development (Antonelli, 2009). In fact, between 1990 and 1997, mining exploration grew by 90 per cent in the world, and 400 per cent in Latin America (Bebbington, 2007, in Svampa and Antonelli, 2009). In Colombia, like in other countries of the region, a new mining law in 2001 granted generous tax exemptions, extended periods for exploitation and diminished administrative requirements for contracts, and made hiring conditions for global corporations more 'flexible' (Kairuz Hernández, 2009; Svampa, Bottaro and Antonelli, 2009). Between 2002 and 2006, mineral exports and mineral production as part of the GDP grew significantly[11] and, unsurprisingly, forty companies requested licenses to exploit gold within more than four million hectares between 2007 and 2009 (Kairuz Hernández, 2009). Unlike traditional mining, the 'new minerals' are not concentrated in veins within mines, but instead scattered in mountains. Therefore, satellite-identified areas are dynamited, and the resulting blown rocks are mixed with chemicals like cyanide and mercury to dissolve the minerals. These technologies are highly wasteful and polluting of water: they use massive amounts of water and leave them untreated, contaminated with these chemicals (Svampa and Antonelli, 2009).

Extractive industries create no bonds with the communities or the places in which they are located. Large-scale mining extracts and exhausts natural resources, to later withdraw in search of other places to exploit, after having wasted and polluted massive amounts of water (Giarracca and Hadad, 2009). This juggernaut of nature and culture is what environmentalists had in mind when prioritizing, in the text of the referendum, the health of ecosystems and the cycle of water, and when referring to 'the territory' – an understanding of the socio-natural world, or the relation between land, resources, people, and wealth, radically different from what extractive industries re-enact, which I elaborate upon below. The reference to water as essential for the life of all living beings and all species, not just humans, resonates with the 2008 Ecuadorean constitutional reform – which content inspired Colombians. In an epistemic-political rupture with key tenets of liberalism, capitalism and the state, the new Ecuadorean Constitution absorbed the indigenous understanding of *pachamama*, or nature, and made it a subject of rights on par with humans (Escobar, 2010). Colombian activists, attuned to this knowledge, defined water beyond the modern understanding of 'natural resources', or inert objects for humans to appropriate, and envisioned it instead as a common and public good, and a carrier of cultural and sacred values. While prioritizing the health of ecosystems, and the water needs of humans and non-humans alike, the text makes them both political actors. The text thus conjures up the entire range of the living beyond humans, echoing non-dualistic, relational and hybrid understandings of social nature – prevalent within critical human geographers' debates in the last two decades (Castree and Braun, 2001; Swyngedouw, 2004).

A pedagogic exercise in the territories

On June 8, 2010, I asked Diego, the environmentalist from Bogotá, if he was surprised with Congress's final rejection of the referendum. He said:

I was not surprised at all. I said on every stage where I could possibly speak that the *referendum was a pedagogic exercise within the territories, and a mobilization tool. Our struggle was not a change in the Constitution* – because constitutions change, and what? Look at the 1991 Colombian Constitution, one of the most progressive in the world for indigenous peoples rights – and indigenous groups are about to disappear and be exterminated in this country. . . . I always said the referendum was an exercise, and we should find the ways of radical democracy within it.

For activists like Diego, the referendum became a tool with which to intervene, and empower the idea and materiality of the territories. The concept of territory emerged in the 1980s when black peasant communities of the Atrato River in the Pacific basin began to strategize against predatory timber companies. They inaugurated a conceptual framework and new type of property that they developed further, in the context of the 1991 constitutional reform and the regulation called Ley 70, as they became the movement now known as PCN. As a key step towards the collective titling of land, PCN activists and members of black and indigenous communities created their own maps. They travelled along their territory to distinguish places of habitation, cultivation and socializing, past and present. Together with oral histories, collective maps narrated use-spaces, production systems, the history of each settlement, local knowledge of plants and animals, and the like. Mapping introduced a new way of representing spatiality, of thinking about the territory, and of expressing the desires of communities (Escobar, 2008, pp55–56). 'The idea that the territory was fundamental to the physical and cultural survival of the communities, and the argument that these communities have unique ways, rooted in culture, of using the diverse spaces constituted by forest, river, mangrove, hills and ocean were the two most important conceptual innovations' (p54). Later on, in workshops and myriad interactions between the river communities, PCN leaders, NGOs, academics and state officials, 'territory came to be defined as the space of effective appropriation of ecosystems by a given community . . .', that is, it evokes 'spaces used to satisfy community needs and to bring about social and cultural development. . . . Thus defined, the territory cuts across several landscape units; more important, it embodies a community's life project' (p146).

'No black person could survive in the Pacific without water', Samuel, the displaced PCN leader told me in Buenaventura, after he proudly invited me to Palenque el Congal (a key meeting place of PCN) to 'go see the map', and while explaining to me why the PCN movement had joined the referendum process. What he meant was obviously not mere physical survival. What he was evoking was a sense of individual and collective self, spiritually and materially anchored in a place, a 'territory that is structured by water', as Diego put it to me. When Samuel introduced himself to me, he spontaneously offered, like many in the Colombian Pacific do, a lot of information about the river where he had been born. He spoke enthusiastically about 'the Anchicayá'. He elaborated on its history since the Spanish conquest. He mentioned the number of streams, cliffs and mangrove swamps. He was eager to discuss the dam located at the headwater which produces power, but not for his own community. As he talked about 'our body' and 'how it is made of water', it was not easy for me to decode when he was referring to the territory and when to his own, individual, human body. The collective, place-based identity that Samuel endeavours to nurture, was not ancestrally handed down to him. It is a materially rooted, symbolic meaning that was organized as such during the 1991 constitutional process, and the collective titling of land. And Samuel, within the PCN movement, tries to safeguard it when-ever possible. Hence, when he elaborated on the referendum campaign, the spiritual and aesthetic values that his community attaches to water matched the importance of the materiality and the economic value of the river. 'We worked in the referendum from the perspective of defending life, because water is life, it is a sacred element, and an element of joy, hope, cleanliness, beauty, and health. And the river is a means of communication and a means of work', Samuel told me on June 19, 2010. And even if he acknowledged that PCN activists supported the leftist party Polo (that endorsed the refer-endum), Samuel emphasized that they had their *own* perspective, and their *own political project* in mind.

After that alternative ordering of the territories in the Pacific region in the 1990s, paramilitaries have been arriving en masse since 2000 – with terror-izing strategies and seeking highly desired land for coca and African oil palm (Escobar, 2008) for the production of cocaine and bio-fuels respectively. For Afro-Colombian and indigenous communities of the region, and for activists supporting them and the conceptual and political innovation of the territory, engaging in the referendum process has become a way of defending a col-lective life project, and reinforcing a 'subaltern strategy of localization'.[12] Beyond the letter of the text, which explicitly defends waters within collective territories as an integral part of them, the 'referendum on water' has become another – safe – way of nourishing the conversation on development, and empowering victimized communities. In the most traumatized places, where people live daily with paramilitaries who patrol them day and night and 'are aware of everything that goes on', both activists and residents need to be ever ready with reasons to justify their whereabouts or their actions, however trivial, when stopped and questioned by paramilitaries (Madariaga Villegas, 2006, p47, my translation). For activists, the referendum could serve as this innocuous excuse; it was something the paramilitaries did not quite understand, much less ascribe any importance to. Diego told me, for example, that in the Cauca region, the paramilitaries did not see the topic as 'something related to guerrillas', and disregarded it as 'simply signatures for something popular, something about the constitution that will never happen'.

The referendum thus allowed activists to nourish a critical consciousness within subaltern communities by posing questions to its members such as, again in the words of Diego, who worked in the Cauca region, 'Who owns biofuel enterprises?' 'Why are people being displaced here?' 'What are the interests behind displacement?' 'Who controls mining?' 'Who controls the diversion of water?' 'What types of wealth are being produced here?' The referendum thus became a security shell for activists, as well as an opportunity for strategic talk to help make oppressed communities somewhat more resistant to the coercive biophysical and cultural reconfiguration wrought by global capitalist development, increasingly connected to the use of terror and armed actors.

An engaged universal

On July 28, 2010, the UN General Assembly recognized access to water and sanitation as a fundamental human right. Roughly two months later, the Geneva-based Human Rights Council (the main UN body on such matters) confirmed that decision. The recently established human right to water, however, in the words of the UN Independent Expert Catarina de Albuquerque:

> do[es] not favour a particular model of service provision and the decision to delegate lies with the State. [. . .] Traditionally, human rights are concerned with the relationship between the State and the individual, imposing obligations on States and endowing individuals with rights. When a third party becomes involved in the realization of human rights, that relationship becomes even more complex.[13]

The reason it becomes even more complex, the Independent Expert reported to the Human Rights Council, is because when water supply is managed by non-state actors, the lack of democratic decision-making, the power asymmetries in the bidding process and the plight of the poorest and most marginalized, among other factors, might jeopardize the universality of human rights.[14]

Bakker (2007, and in this book) cautioned us that the human right to *access* water does not automatically define the character of water as a non-commodity, and thus does not foreclose water provision by private corporations. Indeed, the 2010 universal declaration, even if a mark of progress towards global water justice, did not cancel the private property regime of water provision. Commodification, which activists reject so vehemently, is now compatible with the also vehemently sought after human right to water. Were Latin American and Colombian water warriors mistaken when framing their struggle around the human right to water? Did they misuse their political energies and waste their contentious efforts?

Water struggles are place-based and rooted in particular historical geographies, political economies and cultural contexts. Yet, I have argued here, since the 2000 water war in Cochabamba, there has been an emerging activist repertoire that travels along transnational public spheres like the WSF, Red VIDA workshops and the World Water Forum counter-summits. In such journeys, the repertoire intertwines localities and builds networks of local and global, activists and analysts. Despite the hesitation of some, amending national constitutions to include the access to water as a fundamental human right is a key piece of this travelling repertoire. It was successful in Uruguay in 2004 and in Bolivia in 2009. In Colombia, however, it is far from being materialized. But it became, instead, 'a universal that moves', a mobile and mobilizing tool that engaged and charged particular struggles from below – to borrow from Tsing (2005, pp7–8). In Colombia, where grassroots movements and leaders are terrorized, the referendum on the human right to water became a tool for network building, and a resource to nurture a socio-environmental imagination that challenges development based on global capital, extractive industries and the terror of the armed struggle.

Few local critics have questioned whether the referendum was the 'best choice' for Colombia. Its political culture is strikingly different from that of Uruguay. The latter has a long trajectory of citizens exercising mechanisms of direct democracy, and the notable precedent of the 1992 plebiscite that banned the privatization of all public services. The 2004 Uruguayan referendum on water was voted on the same day that the leftist Frente Amplio won the presidential elections. The overall political climate thus favoured anti-privatization policy reform. In Colombia, on the contrary, there was no history of referendums before the mobilization for water, and the overall milieu is typically described by activists as 'righticized' (*derechoso y derechizado*), and often welcoming of privatization reform. In addition, 'In such "a legalistic country" (*un país tan leguleyo*) with such environmentally progressive rights after 1991', a public intellectual shared his doubts with me, 'is constitutional reform the most effective way to accomplish water justice?' 'Furthermore', he said, 'isn't the legal strategy depoliticizing the social movement and shutting down a deeper national conversation on water?'

All in all, I argue, water warriors in Colombia adopted and adapted a script that emerged in Cochabamba when networked citizens were powerful enough to cancel a privatization contract and evict a United States-led transnational corporation. In this process, I claim, Colombians opened up new spaces for democratic participation. They expanded the socio-environmental stakes and the emancipatory potential of the struggle for the right to water. Today, 'water privatization' does not only mean for-profit water supply by transnational corporations. Water privatization does refer to the political economic process that Harvey (2003) calls accumulation by dispossession. Yet, because it also denounces the pollution, depletion and diversion of water *sources*, that in Colombia often goes hand in hand with the (paramilitary) use of terror to displace local communities, 'water privatization' also works discursively to challenge a capitalist development model rooted in extractive industries like large-scale mining, mega infrastructure and global capital. And because Colombian activists used the referendum as a mobilizing tool to intervene, and empower the idea and materiality of the territories, the struggle also worked to protect the collectives and political process that Afro-Colombians and indigenes began around the 1991 constitutional reform.

Notes

1 VIDA means 'life' in Spanish, and is also the acronym for Vigilancia Interamericana para la Defensa y el Derecho al Agua (Inter-American Watch for the Defense and Right to Water), (Red Vida, 2003). Since its inception in 2003, it has been the most active network on this issue, gathering organizations from fourteen countries in the Americas that struggle against privatization and for public and community ownership and management of water supply services, with citizens' participation.

2 Like all the names I use in this chapter, Diego is a pseudonym.

3 In January 2009, I participated in the global meeting of the World Social Forum, focusing on the workshops, talks and art performances organized by global water activists. In June 2010, thanks to the Robert O. Fehr Professorship from State University of New York Purchase, I travelled to Bogotá, Medellín, Cali, Buenaventura and Juanchaco in the Pacific Basin in Colombia. I participated in national assemblies, forums, and meetings that ranged from ten to two hundred activists, and interviewed in-depth twenty leaders and activists of the referendum process.

4 Bakker (2007, and in this book) argues that 'neoliberalization' of nature is the way critics frame 'market environmentalism' and underscores its lack of clarity. Privatization, marketization, commercialization, deregulation and re-regulation are different policies but often lumped under the same label of neoliberalization. I use the term because I think it is evocative to identify and criticize the overall market inclination of reforms. Yet, in each case, I specify what type of reform I am referring to.

5 Even if the Cochabamba water war was an important reason for the Banzer administration's state of siege, it was not the only one. Other important protests were taking place simultaneously (Assies, 2003).

6 They had to pay seven times more for drinking water, forty times more for sanitary services and eighty times more for new sanitary connections. Also, along with public taps, private companies ended prior 'crossed subsidies' that had made wealthier households from Maldonado County support poorer areas in the country (Santos, 2006).

7 Even if mentioned in the 1934 and 1942 Uruguayan Constitution, plebiscites and referendums became a clear possibility with the 1967 Constitution. Until the 1990s, however, plebiscites were mainly used by party actors to legitimize their decisions (Valdomir, 2006).

8 This account is based on several interviews and information from Ecofondo, available at www.ecofondo.org.co/ecofondo/downloads/3.%20CAMPANA%20 DEL%20AGUA1.pdf, accessed 10 July 2010.

9 The letter is available at www.censat.org/noticias/2009/5/20/Carta-abierta-al-Presidente-de-la-Camara-de-Representantes-y-el-gobierno-de-Alvaro-Uribe/, accessed 26 July 2010.
10 The referendum text is available at www.ecofondo.org/index.php?option=com_content&view=article&id=193:texto-del-articulado-presentado-a-la-registraduria-nacional&catid=45:todo-sobre-el-referendo&Itemid=62, accessed 17 February 2011.
11 Duque Montoya, González Castrillón and Calle, 2006, available at www.simco.gov.co/simco/portals/0/archivos/PROMOCION.pdf, accessed 10 July 2010.
12 Escobar refers to the territory as a 'subaltern strategy of localization' (2008, p59).
13 United Nations Human Rights, 15 September, 2010, available at http://www.ohchr.org/EN/NewsEvents/Pages/DisplayNews.aspx?NewsID=10356&LangID=E, accessed 22 November 2010.
14 United Nations Human Rights, 15 September, 2010, available at http://www.ohchr.org/EN/NewsEvents/Pages/DisplayNews.aspx?NewsID=10356&LangID=E, accessed 22 November 2010.

References

Antonelli, M. (2009) 'Minería transnacional y dispositivos de intervención en la cultura. La gestión del paradigma hegemónico de la "minería responsible y el desarrollo sustentable"', in M. Svampa and M. Antonelli (eds) *Minería transnacional, narrativas del desarrollo y resistencias sociales*, Biblios, Buenos Aires

Assies, W. (2003) 'David versus Goliath in Cochabamba: water rights, neoliberalism and the revival of social protest in Bolivia', *Latin American Perspectives*, vol 30, no 3, pp14–36

Bakker, K. (2007) 'The "commons" versus the "commodity": alter-globalization, anti-privatization and the human right to water in the global South', *Antipode*, vol 39, no 3, pp430–455

Barlow, M. and Clarke, T. (2004) *Blue Gold. The Fight to Stop the Corporate Theft of the World's Water*, The New Press, New York

Boff, L. (1996) *Ecología. Grito de la tierra. Grito de los pobres*, Editorial Trotta, Madrid

Castree, N. and Braun, B. (2001) *Social Nature. Theory, Practice and Politics*, Blackwell Publishers, Malden, MA and Oxford

Duque Montoya, B., González Castrillón, E. and Calle, L. (2006) 'Política de Promoción del País Minero', www.simco.gov.co/simco/portals/0/archivos/PROMOCION.pdf, accessed 18 November 2010

Escobar, A. (2008) *Territories of Difference. Place, Movements, Life, Redes*, Duke University Press, Durham, NC and London

Escobar, A. (2010) 'Latin America at a crossroads', *Cultural Studies*, vol 24, no 1, pp1–65

García Linera, A. (2004) 'The multitude', in O. Olivera (ed) *Cochabamba! Water War in Bolivia*, South End Press, Cambridge, MA

Giarracca, N. and Hadad, G. (2009) 'Disputas manifiestas y latentes en La Rioja minera. Política de vida y agua en el centro de la escena', in M. Svampa and M. Antonelli (eds) *Minería transnacional, narrativas del desarrollo y resistencias sociales*, Biblios, Buenos Aires

Harvey, D. (2003) *The New Imperialism*, Oxford University Press, Oxford

Juris, J. (2008) 'The new digital media and activist networking within anti-corporate globalization movements', in J. Inda and R. Rosaldo (eds) *The Anthropology of Globalization*, Blackwell Publishing, Malden, MA and Oxford

Kairuz Hernández, F. (2009) 'Audiencia Pública en Cajamarca. Para qué?', Diario El Nuevo Día, February 2009, www.biodiversityreporting.org/article.sub?docId= 30193&c=Colombia&cRef=Colombia&year=2009&date=February%202009, accessed 20 November 2010

Madariaga Villegas, P. (2006) *Matan y Matan y Uno Sigue Ahí*. Control paramilitar y vida cotidiana en un pueblo de Urabá, Universidad de los Andes, Facultad de Ciencias Sociales, Bogotá

Olivera, O. (2004) *Cochabamba! Water War in Bolivia*, South End Press, Cambridge, MA

Red Vida (2003) 'VIDA network is born. Newsletter', www.laredvida.org, accessed 21 May 2010

Robertson, R. (1995) 'Glocalization: time-space and homogeneity-heterogeneity', in M. Featherstone, S. Lash and R. Robertson (eds) *Global Modernities*, Sage, London

Salazar Restrepo, B. (2010) *El Agua: Un Derecho Humano Fundamental y un Bien Público*, Corporación Ecológica y Cultural Penca de Sábila, Medellín

Santos, C. (2006) 'La privatización del servicio público de agua en Uruguay', in C. Santos, S. Valdomir, V. Iglesias and D. Renfrew (eds), *Aguas en Movimiento. La resistencia a la privatización del agua en Uruguay*, Ediciones de la Canilla, Montevideo

Schultz, J. (2003) 'Bolivia: the water war widens', *North American Congress on Latin America*, vol. XXXVI, no 4

Svampa, M. and Antonelli, M. (2009) 'Hacia una discussion sobre la megaminería a cielo abierto', in M. Svampa and M. Antonelli (eds) *Minería transnacional, narrativas del desarrollo y resistencias sociales*, Biblos, Buenos Aires

Svampa, M., Bottaro, L. and Antonelli, M. (2009) 'La problemática de la minería metalífera a cielo abierto: modelo de desarrollo, territorio y discursos dominantes', in M. Svampa and M. Antonelli (eds) *Minería transnacional, narrativas del desarrollo y resistencias sociales*, Biblos, Buenos Aires

Swyngedouw, Erik (2004) *Social Power and the Urbanization of Water. Flows of Power*, Oxford University Press, Oxford

Tapia, L. (2000) 'La crisis política de Abril', in *Observatorio Social de América Latina/ Septiembre 2000*, CLACSO, Buenos Aires

Tarrow, S. (2005) *The New Transnational Activism*, Cambridge University Press, New York

Tsing, A. (2005) *Friction. An Ethnography of Global Connection*, Princeton University Press, Princeton

Urrea, D. and Camacho, J. (2007) *Agua y Transnacionales en la Costa Caribe*. Documento Censat Agua Viva, www.censat.org/bd/publicaciones/Agua-y-trasnacionales-en-la-Costa-Caribe.2, accessed 23 July 2010

Valdomir, S. (2006) 'Contexto y antecedentes del "Plebiscito del Agua"', in C. Santos, S. Valdomir, V. Iglesias and D. Renfrew (eds) *Aguas en Movimiento. La resistencia a la privatización del agua en Uruguay*, Ediciones de la Canilla, Montevideo

Waterman, P. (1998) *Globalisation, Social Movements and the New Internationalisms*, Mansell, London

INDEX

Page numbers in *Italics* represent tables.
Page numbers in **Bold** represent figures.

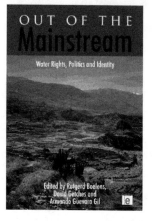

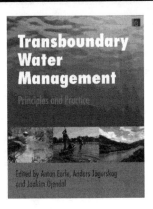

WATER TITLES FROM EARTHSCAN

Water Rights and Social Justice in the Mekong Region

Edited by Kate Lazarus, Nathan Badenoch, Nga Dao and Bernadette P. Resurreccion

The Mekong Region has come to represent many of the important water governance challenges faced more broadly by mainland Southeast Asia. This book focuses on the diverse social, political and cultural dynamics that shape the various realities and scales of water governance in the region, in an effort to bring to the forefront some of the local nuances required in the formulation of a larger vision of justice in water governance. It is hoped that this contextualized analysis will deepen our understanding of the potential of, and constraints on water rights in the region, particularly in relation to the need to realize social justice. The authors show how vitally important it is that water governance is democratized to allow a more equitable sharing of water resources and counteract the pressures of economic growth that may pose risks to social welfare and environmental sustainability.

February 2011: 272pp

Hb: 978-1-84971-188-3: **£60.00**

To Order: Tel: +44 (0) 20 701 76388   **Fax:** +44 (0) 20 7017 6707

or Post: Earthscan, Taylor & Francis Group, Freepost SN926, 2 Park Square, Milton Park, Abingdon, Oxon, OX14 4BR

For a complete listing of all our titles visit:

www.routledge/sustainability.com